10/30/92

DRINKING WATER
HEALTH ADVISORY

VOLATILE
ORGANIC
COMPOUNDS

United States
Environmental Protection Agency
Office of Drinking Water
Health Advisories

Library of Congress Cataloging-in-Publication Data

Drinking water health advisory. Volatile organic compounds.
p.   cm. — (United States Environmental Protection Agency,
Office of Drinking Water health advisories)
    Includes bibliographical references.
    1. Drinking water—Health aspects—United States. 2. Organic
water pollutants—United States—Physiological effect. 3. Drinking
water—Standards—United States. 4. Water—United States—
Purification—Organic compounds removal. I. United States. Environmental Protection Agency. Office of Drinking Water. II. Series.
RA592.A1D76   1991
363.6′1—dc20   90–13323
ISBN 0–87371–436–9

LEWIS PUBLISHERS, INC.
121 South Main Street, Chelsea, Michigan 48118

PRINTED IN THE UNITED STATES OF AMERICA

Preface

Scope and Purpose of the Health Advisory Program

The United States Environmental Protection Agency (U.S. EPA) Office of Drinking Water (ODW) Health Advisory Program was initiated to provide information and guidance to individuals or agencies concerned with potential risk from drinking water contaminants for which no national regulations currently exist. The Health Advisories (HAs) discussed in this volume were developed in a cooperative effort with the Office of Research and Development. HAs are prepared for contaminants that have the potential to cause adverse human health effects and are known or anticipated to occur in drinking water. Each HA contains information on the nature of the adverse health effects associated with the contaminant and the concentrations of the contaminant that would not be anticipated to cause an adverse effect following various periods of exposure. In addition, the HA summarizes information on occurrence, environmental fate, available analytical methods, and treatment techniques for the contaminant.

History and Present Status

The program was begun in 1978, and guidance was issued for the first 20 contaminants in 1979. At that time, the concentrations judged to be safe were termed "Suggested-No-Adverse-Response Levels" (SNARLs). These guidance values were retitled Health Advisories in 1981. To date, the U.S. EPA has issued 119 HAs in final form, covering a wide variety of inorganic, pesticide and nonpesticide and volatile organic contaminants, and one microbial contaminant (*Legionella*). In addition, the U.S. EPA is preparing additional HAs on various organic chemicals, disinfectants and their by-products, and other inorganic contaminants.

Quality Assurance

Initial drafts of each HA undergo a series of thorough reviews before they are released to the public. The general technical content and the risk assessment values are reviewed by a group of independent expert scientists, an ODW Toxicological Review Panel, and other U.S. EPA offices with interest and expertise in the contaminant. The draft HA may also be distributed for review and comment by the public. Each HA is revised in response to criticisms and suggestions received during the review process before being released in final draft form. Each HA is updated as significant new information becomes available that may impact the original conclusions or guidance values.

Acknowledgments

The development of each HA involves the participation of many individuals. The following members of the Health Advisory Program are acknowledged for their valuable contributions:

Michael B. Cook, Director, Office of Drinking Water

Steve Clark, Acting Chief, Science and Technology Branch,
Office of Drinking Water

Margaret Stasikowski, M.S., Director, Criteria and Standards Division,
Office of Drinking Water

Edward V. Ohanian, Ph.D., Chief, Health Effects Branch, Office of
Drinking Water

Jennifer Orme, M.S., Health Advisory Program Manager

Cindy Sonich-Mullin, M.S., Chief, Systemic Toxicants Assessment Branch,
Environmental Criteria and Assessment Office

Charles Abernathy, Ph.D.	Julie Du, Ph.D.
Vyto A. Adomaitis, M.S.	David Huber, M.S.
Larry Anderson, Ph.D.	Krishan Khanna, Ph.D.
Ken Bailey, Ph.D.	Amal Mahfouz, Ph.D.
James Barron, B.A.	Bruce Mintz, B.S.
Robert Cantilli, M.S.	James Murphy, Ph.D.
Nancy Chiu, Ph.D.	Yogendra Patel, Ph.D.

The members of the Health Advisory Program would like to acknowledge the assistance provided by Eastern Research Group, Inc.

Introduction

Health Advisories (HAs) are prepared by the Criteria and Standards Division, Office of Drinking Water (ODW) of the United States Environmental Protection Agency (U.S. EPA) in Washington, DC. Documents summarized in this volume are part of the Health Advisory Program sponsored by ODW in response to the public need for guidance during emergency situations involving drinking water contamination. They provide technical guidance to public health officials on health effects, analytical methodologies, and treatment technologies associated with drinking water contamination.

The HAs for 15 unregulated volatile organic chemicals were developed jointly by the Environmental Criteria and Assessment Office in the Office of Research and Development and ODW. Each HA summarizes available data concerning the occurrence, pharmacokinetics, and health effects of a specific contaminant or mixture, as well as analytical methods and treatment technologies for the contaminant. The health effects data are used to estimate concentrations of the contaminant in drinking water that are not anticipated to cause any adverse noncarcinogenic health effects over specific exposure durations (see Table I). These HA concentrations include a margin of safety to protect sensitive members of the population (e.g., children, the elderly, pregnant women). Health Advisories are used only for guidance and are not legally enforceable in the United States. They are subject to change as new information becomes available.

The data for each HA were obtained through a comprehensive literature search covering available publications through September 1989. Section 1445 of the Safe Drinking Water Act as amended in 1980 requires the U.S. EPA to require public water suppliers to conduct a monitoring program for unregulated contaminants. Each system must monitor at least once every five years for unregulated contaminants. These data will assist U.S. EPA in determining whether regulations are necessary. Fifty-one compounds have been included in the monitoring list. Health Advisories have been drafted for several of these contaminants. This volume includes new Health Advisories for 15 of the unregulated contaminants for which guidance had not previously been available. The HAs provide guidance on health effects and water treatment, should an unregulated contaminant be detected.

Table I. Drinking Water Health Advisories (HAs) Determined by the
Office of Water

- *One-day HA* — The concentration of a chemical in drinking water that is not expected to cause any adverse noncarcinogenic effects for up to 5 consecutive days of exposure, with a margin of safety.

- *Ten-day HA* — The concentration of a chemical in drinking water that is not expected to cause any adverse noncarcinogenic effects up to 14 consecutive days of exposure, with a margin of safety.

- *Longer-term HA* — The concentration of a chemical in drinking water that is not expected to cause any adverse noncarcinogenic effects up to approximately 7 yr (10% of an individual's lifetime) of exposure, with a margin of safety.

- *Lifetime HA* — The concentration of a chemical in drinking water that is not expected to cause any adverse noncarcinogenic effects over a lifetime of exposure, with a margin of safety.

For further information on the Health Advisories, contact the Safe Drinking Water Hotline (800–426–4791); in Alaska or in the Washington, DC area call (202-382–5533) between 8:30 and 4:30 EST M-F or the Health Effects Branch, Criteria and Standards Division, Office of Drinking Water, U.S. EPA, 401 M Street SW, Washington, DC 20460 (202-382–7571).

I. Assessment of Noncarcinogenic Risks

A. Selection of Data for Deriving Health Advisories

Health Advisories are based on data from animal or human studies of acceptable design. The first step in deriving HAs is a thorough review of the literature. For each study, the highest doses at which no adverse effects were observed in the test species (No-Observed-Adverse-Effect Levels — NOAELs) and the lowest doses at which adverse effects were observed (Lowest-Observed-Adverse-Effect Levels — LOAELs) are noted. For each HA, the most appropriate NOAEL (or LOAEL, if a NOAEL has not been identified) is selected from the available data based on the considerations described below.

A key factor in determining which NOAEL or LOAEL to use in calculating a particular HA is exposure duration. Ideally, the data will be taken from a study with an exposure duration comparable to the exposure duration for which the HA is being derived (see Table II). For example, a One-day HA is generally based on data from acute human or animal studies involving up to 7 days of exposure; a Ten-day HA is generally based on subacute animal studies involving 7 to 30 days of exposure.

Another factor that is considered in selecting the NOAEL or LOAEL is the route of exposure. An oral route (drinking water, gavage or diet) is preferred. Data from inhalation studies may also be used in deriving a HA when adequate ingestion data are not available. The relevance of data from subcutaneous or intraperitoneal studies is considered on a case-by-case basis.

Other factors that contribute to selection of the NOAEL or LOAEL are species sensitivity and the magnitude of the NOAEL/LOAEL relative to other NOAELs/LOAELs (generally the lowest concentration is used); the degree of confidence in the study; and whether the NOAEL or LOAEL is supported by other dose-response data.

B. Derivation of the One-day, Ten-day and Longer-term Health Advisories

Once the NOAEL or LOAEL has been selected, the One-day, Ten-day, and Longer-term HAs are derived using the following formula:

$$ HA = \frac{(NOAEL \text{ or } LOAEL) \ (BW)}{(UF) \ (\underline{\quad} \ L/d)} = \underline{\quad} \ mg/L \ (or \ \underline{\quad} \ \mu g/L) $$

where: NOAEL = No-Observed-Adverse-Effect Level
                or   or
            LOAEL = Lowest-Observed-Adverse-Effect Level
               BW = assumed body weight of protected individual
               UF = uncertainty factor (chosen in accordance with EPA or National Academy of Science [NAS] guidelines discussed below)
          __ L/d = assumed daily water consumption of protected individual

The assumptions made concerning human body weights and water consumptions used in calculating each HA are given in Table II. For the One-day and Ten-day HAs, the protected individual is assumed to be a 10-kg child drinking 1 L/d of water. For the Lifetime HA, the protected individual is assumed to be a 70-kg adult consuming 2 L/d of water unless otherwise indicated. Longer-term HAs are calculated for both the child and the adult.

The uncertainty factor, chosen in accordance with NAS/ODW guidelines (Table III), ranges from 1 to 10,000 depending on the nature and quality of the data. Selection of the uncertainty factor is based principally upon scientific judgment and accounts for possible intra- and interspecies differences. Other considerations may necessitate the use of an additional uncertainty factor of 1 to 10. These considerations include the significance of the adverse health effect, pharmacokinetic factors, counterbalancing of beneficial effects, and the quality of the available data base for each contaminant.

Table II. Data Used to Develop Health Advisories (HAs) and Carcinogen Risk Estimates

	Assumed Weight of Protected Individual	Assumed Volume of Drinking Water Ingested/Day	Preferred Exposure Data for HA Development
One-day HA	10-kg child	1 L	Up to 7 days of exposure
Ten-day HA	10-kg child	1 L	7 to 30 days of exposure
Longer-term HA	10-kg child and	1 L	Subchronic (90 days to 1 yr)
	70-kg adult	2 L	(i.e., approximately 10% of the animal's lifetime)
Lifetime HA	70-kg adult	2 L	Chronic or subchronic
Cancer risk estimates	70-kg adult	2 L/day for 70 yrs	Chronic or subchronic

C. Derivation of the Lifetime Health Advisory

The One-day, Ten-day, and Longer-term HAs are based on the assumption that all exposure to the chemical comes from drinking water. Over a lifetime, however, other sources (e.g., food, air) may provide significant additional

Table III. Guidelines Used in Selecting Uncertainty Factors for HAs[a]

• An uncertainty factor of 10 is generally used when good chronic or subchronic human exposure data identifying a NOAEL are available, and are supported by chronic or subchronic toxicity data in other species.

• An uncertainty factor of 100 is generally used when good chronic toxicity data identifying a NOAEL are available for one or more animal species (and human data are not available), or when good chronic or subchronic toxicity data identifying a LOAEL in humans are available.

• An uncertainty factor of 1,000 is generally used when limited or incomplete chronic or subchronic toxicity data are available, or when good chronic data that identify a LOAEL but not a NOAEL for one or more animal species are available.

• An uncertainty factor of 10,000 may be used when a subchronic study identifying a LOAEL but not a NOAEL is used.

[a]Source: NAS (1977, 1980) as modified by the U.S. EPA Office of Drinking Water.

exposure, or may be the predominant exposure route to a chemical. An additional step is added to the calculation of the Lifetime HA to account for these sources.

The Lifetime HA is calculated in three steps. Together, the first two steps are identical to the calculation performed to derive the other HAs. In the first step, the NOAEL or LOAEL is divided by the uncertainty factor to determine a Reference Dose (RfD):

$$\text{RfD} = \frac{\text{(NOAEL or LOAEL)}}{\text{(UF)}} = \underline{\hspace{1cm}} \text{ mg/kg bw/d}$$

The RfD is an estimate, with an uncertainty of perhaps an order of magnitude, of a daily exposure that is likely to be without appreciable risk of deleterious health effects in the human population (including sensitive subgroups) over a lifetime.

In the second step, the RfD is adjusted for an adult (with body weight assumed to be 70 kg) consuming 2 L water per day to produce the Drinking Water Equivalent Level (DWEL):

$$\text{DWEL} = \frac{\text{(RfD) (70 kg)}}{\text{(2 L/d)}} = \underline{\hspace{1cm}} \text{ mg/L } (\underline{\hspace{1cm}} \mu\text{g/L})$$

The DWEL represents the concentration of a substance in drinking water that is not expected to cause any adverse noncarcinogenic health effects in humans over a lifetime of exposure. The DWEL is calculated assuming that all exposure to the chemical comes from drinking water.

In the third step, the Lifetime HA is calculated by reducing the DWEL in proportion to the amount of exposure from drinking water relative to other sources (e.g., food, air). In the absence of actual exposure data, this relative source contribution (RSC) is generally assumed to be 20%. Thus:

$$\text{Lifetime HA} = \text{DWEL} \times \text{RSC} = \underline{\hspace{1cm}} \text{ mg/L } (\underline{\hspace{1cm}} \mu\text{g/L})$$

The value presented in μg/L is generally rounded to one significant figure.

Lifetime HAs are calculated for all noncarcinogenic chemicals (Groups D and E—see Table IV). For known (Group A) or probable (Group B) human carcinogens, carcinogenicity is usually considered the toxic effect of greatest concern. In general, a Lifetime HA is not recommended for Group A or B carcinogens. Instead, a mathematical model (usually the multistage) is used to determine theoretical upper-bound lifetime cancer risks based on the available cancer data (see Section II). For comparison purposes, a DWEL is calculated, and the upper-bound cancer risk associated with lifetime exposure to the DWEL is determined (see Section II).

For chemicals classified in Group C: Possible human carcinogen, ODW applies an additional 10-fold uncertainty factor when deriving a Lifetime HA.

Table IV. EPA Scheme for Categorizing Chemicals According to Their Carcinogenic Potential[a,b]

Group A: Human carcinogen

Sufficient evidence in epidemiologic studies to support causal association between exposure and cancer

Group B: Probable human carcinogen

Limited evidence in epidemiologic studies (Group B1) *and/or* sufficient evidence from animal studies (Group B2)

Group C: Possible human carcinogen

Limited evidence from animal studies *and* inadequate or no data in humans

Group D: Not classifiable

Inadequate or no human *and* animal evidence of carcinogenicity

Group E: No evidence of carcinogenicity for humans

No evidence of carcinogenicity in at least two adequate animal tests in different species *or* in adequate epidemiologic and animal studies

[a]Source: U.S. EPA 1986.
[b]Other factors such as genotoxicity, structure-activity relationships and benign versus malignant tumors may influence classification.

This extra uncertainty factor provides an additional margin of safety to account for the possible carcinogenic effects of the chemical.

II. Assessment of Carcinogenic Risk

If toxicological evidence leads to the classification of a contaminant as a human or probable human carcinogen (Groups A or B), mathematical models are used to estimate an upper-bound excess cancer risk associated with lifetime ingestion of drinking water. The data used in these estimates usually come from lifetime exposure studies in animals. Upper-bound excess cancer risk estimates may be calculated using models such as the one-hit, Weibull, logit, probit, or multistage models (U.S. EPA, 1986). Since the mechanism of cancer is not well understood, there is no evidence to suggest that one model can predict risk more accurately than another. Therefore, the U.S. EPA generally uses one of the more conservative models for its carcinogen risk assessment: the linearized multistage model (U.S. EPA 1986). This model fits linear dose-response curves to low doses (NAS 1986). It is consistent with a no-threshold model of carcinogenesis, i.e., exposure to even a very small amount of the substance theoretically produces a finite increased risk of cancer.

The linearized multistage model uses dose-response data from the most appropriate carcinogenic study to calculate a carcinogenic potency factor (q_1*) for humans. The q_1* is then used to determine the concentrations of the chemical in drinking water that are associated with theoretical upper-bound excess lifetime cancer risks of 10^{-4}, 10^{-5}, and 10^{-6} (i.e., concentrations predicted to contribute an incremental risk of 1 in 10,000, 1 in 100,000, and 1 in 1,000,000 individuals over a lifetime of exposure). The following formula is used for this calculation:

$$\text{Concentration in drinking water} = \frac{(10^{-x})\ (70\ \text{kg})}{(q_1*)\ (2\ \text{L/d})} = \underline{\hspace{1cm}} \ \mu\text{g/L}$$

where: 10^{-x} = risk level (x = 4, 5, or 6)

   70 kg = assumed body weight of adult human

   q_1* = carcinogenic potency factor for humans as determined by the linearized multistage model in $(\mu\text{g/kg/d})^{-1}$

   2 L/d = assumed water consumption of adult human

The carcinogenic risk associated with lifetime exposure to the Drinking Water Equivalent Level (DWEL) is calculated using the following formula:

$$\text{Risk} = (\text{DWEL})\frac{(2\ \text{L/d})\ (q_1*)}{(70\ \text{kg})} = 10^{-x}$$

where: DWEL = Drinking Water Equivalent Level in μg/L

The theoretical upper-bound cancer risk associated with lifetime exposure to the DWEL is provided to assist the risk manager for comparison in assessing the overall risks.

III. Analytical Methods and Treatment Technologies

In addition to the health assessments, HAs also summarize information on analytical methods and treatment technologies for each contaminant. These methods and technologies include those validated as U.S. EPA methods, as well as other methods that may be available. For further information on the analytical methods and treatment technologies for drinking water contaminants, contact the Safe Drinking Water Hotline (800–426–4791) or the Science and Technology Branch, Criteria and Standards Division, Office of Drinking Water, U.S. EPA, 401 M Street SW, Washington, DC 20460 (202–382–3022).

IV. References

NAS. 1977. National Academy of Sciences. Drinking water and health, Vol I. NAS, Washington, DC.

NAS. 1980. National Academy of Sciences. Drinking water and health, Vol II. NAS, Washington, DC.

NAS. 1986. National Academy of Sciences. Drinking water and health, Vol VI. NAS, Washington, DC.

U.S. EPA. 1986. Guidelines for carcinogen risk assessment. *Fed. Reg.* 51(185):33992–34003.

Table of Contents

VOLATILE
ORGANIC
COMPOUNDS

1,1,2-Trichloroethane

I. General Information and Properties

A. CAS No. 79–00–5

B. Structural Formula

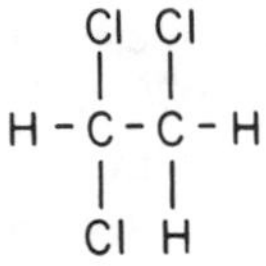

```
      Cl  Cl
      |   |
  H - C - C - H
      |   |
      Cl  H
```

1,1,2-Trichloroethane

C. Synonyms

• Vinyl trichloride, B-trichloroethane, ethane trichloride, 1,1,2-TCE.

D. Uses

• 1,1,2-Trichloroethane is used as a feedstock intermediate in the production of 1,1-dichloroethylene (Archer, 1979); a solvent for chlorinated rubbers (Archer, 1979); and in fats, oils, waxes, and resins (Hawley, 1981).

E. Properties (Sato and Nakajima, 1979; Konemann, 1981; Verschueren, 1983; Windholz et al., 1983; Weast and Astle, 1986)

Chemical Formula	$CH_2ClCHCl_2$
Molecular Weight	133.41
Physical State	Colorless liquid
Boiling Point	113.8° C
Melting Point	-36.5°C
Density	1.44
Vapor Pressure (at 20°C)	19 mm Hg
Specific Gravity	—
Water Solubility	Insoluble (< 20 mg/L)

1

Log Octanol/Water
   Partition Coefficient   2.38 (estimated)
Taste Threshold   –
Odor Threshold   –
Conversion Factor   1 ppm $= 5.55$ mg/m^3
                     1 mg/m^3 $= 0.18$ ppm

F. Occurrence

- Krill and Sonzogni (1986) reported that 1,1,2-trichloroethane was among the most frequently detected volatile organic compounds found in Wisconsin's community and private wells. However, it was never found to exceed 6.1 μg/L.

- 1,1,2-Trichloroethane was detected in 53/603 surface waters from New Jersey during 1977 through 1979 at concentrations ranging from trace to 18.7 ppb. It was also detected in 72/1,069 ground waters at concentrations ranging from trace to 31.1 ppb (Page, 1981).

- Concentrations of 1,1,2-trichloroethane in finished water from the New Orleans, LA, area ranged from 0.1 to 8.5 μg/L (U.S. EPA, 1975).

G. Environmental Fate

- Volatilization is the primary transport process for 1,1,2-trichloroethane in water (Callahan et al., 1979). The half-life for evaporation from a 1 ppm aqueous solution 6.5 cm deep was experimentally determined (25°C, 200 rpm stirring in still air) to be 35.1 minutes (Dilling, 1977). Zoeteman et al. (1980), however, estimated the overall half-life of 1,1,2-trichloroethane in Netherlands surface water to be 1.9 days.

- 1,1,2-Trichloroethane is not degraded in sandy soil, and it percolates rapidly through the soil (Wilson et al., 1981).

- Only slow biodegradation of 1,1,2-trichloroethane was demonstrated in a static culture flask test with domestic wastewater inoculum (Tabak et al., 1981); degradation was 6% after 7 days and 39% after 28 days when the initial concentration of 1,1,2-trichloroethane was 10 mg/L. When the initial concentration was 5 mg/L, degradation was 0% and 44% after 7 and 28 days, respectively.

- There was no detectable degradation of 1,1,2-trichloroethane in ground water (Wilson et al., 1983).

- Chemical degradation (photolysis, oxidation and hydrolysis) of 1,1,2-trichloroethane in water is not expected to be significant (U.S. EPA, 1983).

II. Pharmacokinetics

A. Absorption

- Mitoma et al. (1985) administered single gavage doses of ^{14}C-1,1,2-trichloroethane in corn oil at 70 mg/kg to male Osborne-Mendel rats and 300 mg/kg to male B6C3F$_1$ mice after oral administration of the same dose of the unlabeled compound to the rats and mice 5 days/week for 4 weeks. For 2 days after administration of the radioactive compound, expired gases and urine, but not feces, were collected and radioactivity was determined. For rats, expired gases contained 14.6%, urine contained 72% and the carcass contained 3.9% of the applied dose of radioactivity. For mice, these compartments contained 9.9, 76 and 12.3%, respectively, of the applied dose of radioactivity. These data indicate that 1,1,2-trichloroethane is extensively absorbed from the gastrointestinal tract.

- A single application of 1 mL of 1,1,2-trichloroethane to the skin of guinea pigs was absorbed rapidly as indicated by the presence of 3 to 4 μg/mL of the solvent in the blood in 30 minutes. After 12 hours, the blood concentration increased to almost 5 μg/mL (Jakobson et al., 1977).

B. Distribution

- Little information was found in the available literature regarding the distribution of 1,1,2-trichloroethane to individual tissues.

- In mice given ^{14}C-1,1,2-trichloroethane by intraperitoneal (i.p.) injection at 0.1 to 0.2 g/kg, 2.3% of the administered dose remained in the carcass 3 days later (Yllner, 1971).

- In rats and mice given 70 or 300 mg of ^{14}C-1,1,2-trichloroethane/kg, respectively, by gavage, 5 days/week for 4 weeks followed by a single dose of ^{14}C-1,1,2-trichloroethane (Mitoma et al., 1985), carcass levels of radioactivity 2 days after treatment ranged from 2.3 to 3.9% of the administered dose. Covalent binding of the ^{14}C-1,1,2-trichloroethane and/or metabolites to hepatic protein was low in both species.

C. Metabolism

- Yllner (1971) injected female albino mice with 0.1 to 0.2 g ^{14}C-1,1,2-trichloroethane/kg i.p. and collected expired air, urine and feces for 3 days. Major urinary metabolites identified were chloroacetic acid, S-carboxymethylcysteine and thiodiacetic acid. Small amounts of glycolic acid, 2,2-dichloroethanol, oxalic acid, trichloroacetic acid and 2,2,2-

trichloroethanol were also detected. Of the radioactivity in expired air, 91% (8% of the dose) was determined to be unaltered compound. He noted a striking similarity between the urinary metabolic profiles of 1,1,2-trichloroethane and chloroacetic acid, and suggested that the metabolism of 1,1,2-trichloroethane proceeds chiefly via chloroacetate.

- Ikeda and Ohtsuji (1972) exposed young (70 g) male and female Wistar rats (five to six rats/group) to 200 ppm (1,110 mg/m^3) of 1,1,2-trichloroethane vapors for 8 hours. Major urinary metabolites identified during the 48-hour period from the beginning of exposure were total trichloro-compounds, trichloroacetic acid and trichloroethanol. Qualitatively and quantitatively similar results were obtained whether the chemical was given by i.p. injection of a 2.78 mmol/kg (381 mg/kg) dose or through inhalation exposure of 200 ppm.

- Mitoma et al. (1985) orally dosed male adult mice and rats with 300 or 70 mg 1,1,2-trichloroethane/kg/day, respectively [the maximum tolerated doses (MTD) as by NCI, 1978; another group received $1/4$ of the MTD], 5 days/week for 4 weeks followed by a single dose of ^{14}C-1,1,2-trichloroethane. Based on the sum of the radioactivity located in expired carbon dioxide, excreta and the carcass 48 hours after treatment, the authors determined that rats and mice metabolized approximately 80% of their respective doses. The urine contained 72 to 76% of the dose of radioactivity in both species and the pattern of urinary excretion was qualitatively similar for both species. The amount of 1,1,2-trichloroethane metabolized by each species was proportional to the dose administered, expressed on a mmol/kg basis.

- Takano et al. (1985) observed that 1,1,2-trichloroethane was less tightly bound to male rat liver cytochrome P-450 than 1,1,1-trichloroethane, although its rate of metabolism was greater.

- Chang et al. (1985) showed that 1,1,2-trichloroethane was cytotoxic to primary cultures of rat hepatocytes. Treatment with SKF-525A inhibited cytochrome P-450 oxidase activity, but enhanced the cytotoxicity, suggesting that the metabolism noted by Takano et al. (1985) was a detoxification pathway for 1,1,2-trichloroethane.

- Thompson et al. (1984, 1985) showed that 1,1,2-trichloroethane may also undergo *in vitro* and *in vivo* reductive metabolism to vinyl chloride, in a process mediated by cytochrome P-450 and inhibited by oxygen. This route of metabolism is relatively slow, due to the large negative half-wave reduction potential of -2.33V (Thompson et al., 1984).

- *In vitro* studies indicate that metabolism appears to involve dechlorination by an enzyme system that is similar to a hepatic microsomal mixed-function oxidase system, but also requires a factor from the soluble portion of the liver homogenate (Van Dyke and Wineman, 1971).

D. Excretion

- Yllner (1971) injected mice intraperitoneally with 0.1 to 0.2 g ^{14}C-1,1,2-trichloroethane/kg and measured 73 to 87% of the dose of radioactivity in the urine, 0.1 to 2% in the feces (contaminated by urine) and 16 to 22% in expired air over a 3-day period. Approximately 1 to 3% remained in the carcass.

- Two days after a single oral administration of ^{14}C-1,1,2-trichloroethane to mice and rats, 72 to 76% of radioactivity was excreted in the urine, 10 to 15% in expired air and 2 to 4% remained in the carcass (Mitoma et al., 1985).

III. Health Effects

A. Humans

1. Short-term Exposure

- Grant (1974) reported that exposure to 1,1,2-trichloroethane vapors irritated the respiratory tract and eyes, but recovery occurred within 48 hours after exposure.

- Hardie (1964) reported that 1,1,2-trichloroethane had a narcotic action.

- Wahlberg (1984a) exposed the volar region of the forearm of human subjects to 1.5 mL of 1,1,2-trichloroethane for 5 minutes and observed 100% increases in blood flow to the skin as measured with a laser Doppler flow meter. No erythema was observed. The subjects reported a burning and/or stinging sensation after exposure and spontaneous transient whitening of the skin was observed as well. Maximum blood flow was reached within 10 minutes after administration, and had returned to normal 60 minutes after exposure.

- Wahlberg (1984b) applied 0.1 mL of 1,1,2-trichloroethane once daily to the volar forearm skin of human subjects for 15 days. No persistent erythema or increased skin-fold thickness was observed.

2. Long-term Exposure

- Hardie (1964) reported that long-term exposure to 1,1,2-trichloroethane vapors resulted in symptoms of chronic gastritis, fat deposition in the kidneys and damage to the lungs.

B. Animals

1. Short-term Exposure

- Tyson et al. (1983) fasted three to five adult male Sprague-Dawley rats before administering single doses of 1,1,2-trichloroethane ranging from 0.0 to 1.75 mmol/kg bw (0.0 to 233 mg/kg) in corn oil by gavage. Hepatotoxicity was indicated by increased serum glutamic oxalacetic transaminase (SGOT) and serum glutamic pyruvic transaminase (SGPT) activities, which rose to maximum levels 48 hours after exposure. The dose at which SGOT and SGPT were elevated was 0.45 mmol/kg bw (60 mg/kg bw), which would be a LOAEL for the hepatotoxic effects of 1,1,2-trichloroethane in these experiments.

- White et al. (1985) reported oral LD_{50}s of 378 and 491 mg/kg bw in male and female CD-1 mice, respectively. Target organs affected were the upper GI tract, the liver and the lung.

- Lundberg et al. (1986) exposed female Sprague-Dawley rats to an i.p. injection of 1,1,2-trichloroethane in peanut oil vehicle. The 24-hour and 14-day LD_{50} values were 405 and 265 mg/kg bw, respectively. One-eighth of the LD_{50} dose did not elevate the enzyme activity of sorbitol dehydrogenase which was used as an indicator of liver damage.

- Smyth et al. (1969) reported an oral LD_{50} of 0.58 mL/kg bw (835 mg/kg bw) of 1,1,2-trichloroethane in rats.

- In an early study of the anthelmintic effects of 1,1,2-trichloroethane in dogs, Wright and Schaffer (1932) reported fatty degeneration and central necrosis of the liver and inflammation of the GI tract in dogs given single doses of 0.1 to 0.5 mL/kg bw (144 to 720 mg/kg). The 0.5 mL/kg dose (720 mg/kg) was lethal.

- White et al. (1985) and Sanders et al. (1985) treated three groups of ≥ 12 adult male CD-1 mice each by gavage with 1,1,2-trichloroethane (in 10% emulphor) at 0, 3.8 or 38 mg/kg/day for 14 days. Organ weights (brain, liver, spleen, lungs, thymus, kidneys and testes), hematology, serum chemistry and immune function were evaluated. A statistically significant increase in the absolute weights of the brain, thymus and testes was noted at 38 mg/kg/day. When compared on a body weight or brain weight basis, no differences in any organ were evident. A statistically significant decrease in serum lactic dehydrogenase activity, which was depressed 21% compared with controls, was reported in the 38 mg/kg group. Based on these findings, a NOAEL of 3.8 mg/kg/day was identified for both studies.

2. Dermal/Ocular Effects

- Wahlberg (1984b) reported that dermal application of 0.1 mL of 1,1,2-trichloroethane once daily for ≥ 10 days induced edema and erythema in rabbits and guinea pigs.

- Duprat et al. (1976) reported severe irritation of the skin and mild irritation of the eyes in rabbits exposed to a single dermal application or eye instillation of 1,1,2-trichloroethane.

- Smyth et al. (1969) reported a dermal LD_{50} of 3.73 mL/kg (537 mg/kg bw) of 1,1,2-trichloroethane in rabbits.

3. Long-term Exposure

- White et al. (1985) administered 1,1,2-TCE (95% purity) in drinking water to CD-1 mice of both sexes (16/sex) at levels of 20, 200 and 2,000 mg/L for 90 days. Control groups of 24 mice/sex were maintained. Based on fluid consumption and body weight data, average intakes were determined to be 0, 4.4, 46 and 305 mg/kg/day for males and 0, 3.9, 44 and 384 mg/kg/day for females. Parameters of toxicity included fluid consumption, organ weights, hematology, serum chemistry and hepatic microsomal activity. Fluid consumption in high-dose males was depressed 30% and a dose-related reduction in weight gain was statistically significant at this level. High-dose females had reduced hemoglobin and hematocrit values and mice of both sexes had altered serum chemistries that indicated adverse effects on the liver. In the females, these changes, primarily in the high-dose group, included significant increases in liver glutathione, SGOT, SGPT, serum alkaline phosphatase (SAP), fibrinogen levels and liver weight, both absolute and as % body weight. The SGOT, SAP and fibrinogen levels were significantly ($p <$ 0.05) increased at all dose levels, but did not follow a typical dose-response curve. In mid- and high-dose males, the only comparable changes were significant decreases in liver glutathione and in absolute liver and kidney weights and a significant increase in SAP at the high-dose only. While hepatic microsomal activities were not affected at any dose in male mice, mid- and high-dose females had a significant increase in cytochrome P-450 content and aniline hydroxylase activity. In these studies, the Lowest-Observed-Adverse-Effect Level (LOAEL) seen in females was 44 mg/kg/day, which resulted in a reduction of cytochrome P-450 levels and aniline hydroxylase activity and an increase in SGOT, SAP and fibrinogen. In males, the LOAEL was 46 mg/kg/day, based on a reduction of liver glutathione. The NOAEL in males and females was 20 mg/L (4.4 and 3.9 mg/kg/day, respectively).

- Sanders et al. (1985) assessed immunological effects of 1,1,2-TCE on the same CD-1 mice used in the White et al. (1985) study following 90-day exposure to the chemical in drinking water (4.4, 46 and 305 mg/kg for males and 3.9, 44 and 384 mg/kg for females). Humoral immune status was evaluated by enumeration of IgM antibody forming cells (AFC) against sheep erythrocytes (sRBC), measurement of hemagglutination titers and evaluation of spleen lymphocyte responsiveness to lipopoly-saccharides (LPS). Cell-mediated immune status was evaluated by the delayed-type hypersensitivity (DTH) and popliteal lymph node prolifer-ation responses to sRBC. Cell-mediated immunity was unaltered in both sexes by exposure to 1,1,2-TCE at any of the levels used. Humoral immune status, on the other hand, was depressed in mice of both sexes in the mid- and high-dose groups, particularly when determined by hemagglutination titers. Macrophage function was depressed ($p < 0.05$) only in the high-dose males and was indicated by the ability of thioglycolate-recruited peritoneal exudate cells to phagocytize sRBC. Spleen lymphocyte responsiveness to the B cell mitogen, LPS, was unal-tered in males, but was significantly decreased in females exposed to the highest concentration of the 1,1,2-TCE. Females also exhibited a greater degree of hemagglutination depression than males following exposure to 1,1,2-TCE. The authors concluded that there are sex differences in the immunological response of animals to the chemical, which can be seen only when certain assays are employed. A NOAEL of 20 mg/L (4.4 for males and 3.9 mg/kg/day for females) was identified for this study.

- NCI (1978) administered 1,1,2-TCE in corn oil to Osborne-Mendel rats and B6C3F$_1$ mice of both sexes (50/sex) by gavage for 5 days/week for 78 weeks followed by a 13-week (in mice) to 35-week (in rats) observa-tion period. Rats were exposed to Time-weighted Average (TWA) doses of 0, 46 or 92 mg/kg/day. No nonneoplastic, dose-related changes were reported for either sex of either species. However, high-dose treated rats had a hunched appearance, rough fur, urine stains on abdomen, dysp-nea, squinted eyes with reddish exudate and respiratory difficulty in the latter part of the study. Treated mice, given TWA dosages of 195 and 395 mg/kg, did not show any abnormal clinical signs other than abdomi-nal distention subsequently determined to be due to liver tumors (see

Carcinogenicity section). This study establishes a LOAEL of 46 mg/kg/day for rats, based on clinical signs of toxicity.

4. Reproductive Effects

- No data were found in the available literature on the reproductive effects of 1,1,2-trichloroethane.

5. Developmental Effects

- Seidenberg et al. (1986) screened 1,1,2-TCE for developmental toxicity by administering the compound, by gavage in corn oil, to pregnant ICT/SIM mice at a dose level of 350 mg/kg/day on gestation days 8 to 12. Corn oil was used as the vehicle control. The mice were allowed to deliver, and the neonates were examined and weighed on the day of birth (day 1) and day 3. No significant effects on litter size, pup survival or pup weights were observed. Maternal mortality occurred in 3 of 30 cases.

6. Mutagenicity

- Barber et al. (1981) reported that 1,1,2-trichloroethane was not mutagenic in the *Salmonella typhimurium* reverse mutation assay with or without metabolic activation, by standard plate incorporation assay (Ames test) and by exposure to the vapor phase in the closed incubation system. Simmon et al. (1977) also reported negative results in similar tests with *S. typhimurium*.

- Tu et al. (1985) reported weakly positive results for 1,1,2-trichloroethane in the BALB/C-3T3 cell transformation assay. DiRenzo et al. (1982) reported covalent binding of ^{14}C-1,1,2-trichloroethane at 0.35 mmol/mg DNA/hour to calf thymus DNA *in vitro* following bioactivation by hepatic microsomes isolated from phenobarbital-treated rats.

7. Carcinogenicity

- In the NCI (1978) study cited above, no statistically significant increased incidence in tumors at any site was observed in rats. However, a statistically significant ($p < 0.001$), dose-related increase in hepatocellular carcinomas in mice was observed.

- In both male and female mice, a significant increase in the incidence of hepatocellular carcinomas ($p < 0.001$), as shown by both the Fisher exact test and the Cochran-Armitage test, occurred as a result of administration of 1,1,2-TCE. Hepatocellular carcinomas were observed in 2 of 20 (10%) vehicle control males, 18 of 49 (37%) low-dose males and 37 of 49 (76%) high-dose males. In females, hepatocellular carcinomas were seen in 0 of 20 vehicle control mice, 16 of 48 (33%) low-dose mice and 40 of 45 (89%) high-dose mice. Time to first observed hepatocellular carcinomas was also markedly decreased for mice treated with 1,1,2-TCE, with the first tumor observed at 19 and 49 weeks in high-dose males and females, respectively; in vehicle controls, the first tumor was observed at 90 weeks (NCI, 1978).

- A positive dose-related association between 1,1,2-TCE administration and the incidence of adrenal gland pheochromocytomas in mice of both sexes was indicated by the Cochran-Armitage test. Fisher exact tests confirmed these results for high-dose female mice but not for other groups (NCI, 1978). There were no other neoplasms for which statistical tests indicated a positive association between dosage and tumor incidence in mice. It was concluded that under the conditions of the bioassay, 1,1,2-TCE is carcinogenic in B6C3F$_1$ mice, causing hepatocellular carcinomas and adrenal pheochromocytomas (NCI, 1978).

IV. Quantification of Toxicological Effects

A. One-day Health Advisory

The study by Tyson et al. (1983) has been selected to serve as the basis for the 10-kg child One-day HA because it is the only recent oral study (gavage) of appropriate duration (1 day) in which parameters other than lethality were considered. In this study, a single dose of 60 mg/kg/day in rats was considered to be the effective dose producing an increase in serum transaminase above normal levels in 50% of the test animals (ED$_{50}$). The target organ for the effects of acute exposure to 1,1,2-trichloroethane in rats is the liver. The liver also appears to be the target organ in humans (HSDB, 1985), mice (White et al., 1985) and dogs (Wright and Schaffer, 1932) as well. The Tyson et al. (1983) study indicates adverse effects on the liver based on transaminase activity, which is suggestive of a functional change in the liver.

The One-day HA for the 10-kg child is calculated as follows:

$$\text{One-day HA} = \frac{(60 \text{ mg/kg/day}) \times (10 \text{ kg})}{(1,000)(1 \text{ L/day})} = 0.6 \text{ mg/L (600 } \mu\text{g/L)}$$

B. Ten-day Health Advisory

The 14-day range-finding oral toxicity studies of White et al. (1985) and Sanders et al. (1985) in the mouse are the most suitable studies for estimation of the Ten-day HA. In these studies, a dose of 3.8 mg/kg/day of 1,1,2-TCE produced no observable toxic effects and may be considered the NOAEL. On the other hand, 38 mg/kg/day produced significant increases in the weights of several organs.

The Ten-day HA for a 10-kg child is calculated as follows:

$$\text{Ten-day HA} = \frac{(3.8 \text{ mg/kg/day}) \times (10 \text{ kg})}{(100)\ (1 \text{ L/day})} = 0.38 \text{ mg/L (rounded to } 400 \text{ } \mu\text{g/L)}$$

C. Longer-term Health Advisory

The studies by White et al. (1985) and Sanders et al. (1985) were selected as the basis for determination of the Longer-term HA of 1,1,2-TCE. Animals ingested 0, 20, 200 or 2,000 mg 1,1,2-TCE/L of water; corresponding intakes were 0, 4.4, 46 and 305 mg/kg/day for males and 0, 3.9, 44 and 384 mg/kg/day for females. Clinical chemistry parameters and organ weights indicated adverse effects on the liver of mid- and high-dose females and high-dose males. Hematological effects occurred in high-dose female mice only, and depressed humoral immune status was observed in both sexes consuming 200 mg/L (44 mg/kg/day) or 2,000 mg/L (384 mg/kg/day). The NOAEL of 20 mg/L (3.9 mg/kg/day) was chosen as the basis for the Longer-term HA.

The Longer-term HA for the 10-kg child is calculated as follows:

$$\text{Longer-term HA} = \frac{(3.9 \text{ mg/kg/day}) \times (10 \text{ kg})}{(100)\ (1 \text{ L/day})} = 0.39 \text{ mg/L (rounded to } 400 \text{ } \mu\text{g/L)}$$

The Longer-term HA for the 70-kg adult is calculated as follows:

$$\text{Longer-term HA} = \frac{(3.9 \text{ mg/kg/day}) \times (70 \text{ kg})}{(100)\ (2 \text{ L/day})} = 1.37 \text{ mg/L (rounded to } 1,000 \text{ } \mu\text{g/L)}$$

D. Lifetime Health Advisory

Appropriate studies were not found in the available literature to calculate a Lifetime HA. The 90-day mouse studies by White et al. (1985) and Sanders et al. (1985), as described for the Longer-term HA, have been used to determine the Lifetime HA. A NOAEL of 20 mg/L (3.9 mg/kg/day) was identified for female mice. An additional uncertainty factor was used to account for a less-than-lifetime exposure.

Step 1: Determination of the Reference Dose (RfD)

$$\text{RfD} = \frac{(3.9 \text{ mg/kg/day})}{(1,000)} = 0.004 \text{ mg/kg/day (rounded from } 0.0039 \text{ mg/kg/day)}$$

Step 2: Determination of Drinking Water Equivalent Level (DWEL)

$$DWEL = \frac{(0.0039 \text{ mg/kg/day}) \times (70 \text{ kg})}{2 \text{ L/day}} = 0.137 \text{ mg/L (rounded to}$$
$$100 \text{ } \mu\text{g/L)}$$

Step 3: Determination of Lifetime HA

$$\text{Lifetime HA} = \frac{(0.137 \text{ mg/L}) \times (0.20)}{10} = 0.0027 \text{ mg/L (rounded to}$$
$$3 \text{ } \mu\text{g/L)}$$

E. Evaluation of Carcinogenic Potential

- IARC (1979) concluded that there is limited evidence that 1,1,2-TCE is carcinogenic in mice.

- In an NCI (1978) cancer study in rats and mice, 1,1,2-trichloroethane was associated with increased incidence of hepatocellular carcinomas and adrenal tumors in mice. The U.S. EPA (1980) calculated a human q_1^* of 5.73×10^{-2} (mg/kg/day)$^{-1}$ based on the incidence of hepatocellular carcinomas in male mice. The drinking water exposure levels corresponding to excess lifetime cancer risks of 10^{-4}, 10^{-5} and 10^{-6} were estimated to be 61.3, 6.13 and 0.613 μg/L, respectively, with the assumption that a 70-kg adult has a daily water intake of 2 liters.

- Applying the criteria described in U.S. EPA (1986), 1,1,2-TCE may be classified in Group C: possible human carcinogen. This category is for chemicals with limited evidence of carcinogenicity in animals and no human data.

V. Other Criteria, Guidance and Standards

- ACGIH (1986) recommended a Threshold Limit Value (TLV) of 10 ppm (55 mg/m³) and a Short-term Exposure Level (STEL) of 20 ppm (111 mg/m³) for 1,1,2-trichloroethane. OSHA recommended a standard of 10 ppm (55 mg/m³) for 1,1,2-TCE (CFR, 1985).

VI. Analytical Methods

- Analysis of 1,1,2-trichloroethane is by a purge-and-trap gas chromatographic procedure used for the determination of volatile organohalides in drinking water (U.S. EPA, 1985a). This method calls for the bubbling of an inert gas through the sample and trapping volatile compounds on an adsorbent material. The adsorbent material is heated to drive off the compounds onto a gas chromatographic column. The gas chromatograph is temperature-programmed to separate the method analytes, which are then detected by a halogen specific detector. This method is applicable to the measurement of 1,1,2-trichloroethane over a concentration range of 0.07 to 1500 μg/L. Confirmatory analysis is by mass spectrometry (U.S. EPA, 1985b). The detection limit for confirmation by mass spectrometry has not been determined.

VII. Treatment Technologies

- Leighton and Calo (1981) reported experimental measurements of the distribution coefficients for 21 chlorinated hydrocarbons, including 1,1,2-trichloroethane, in a dilute air-water system. (The distribution coefficient is the ratio of the volume of the compound in air to the volume of the compound in water after purging.) They determined the distribution coefficient for 1,1,2-trichloroethane to be approximately 45.5 at 25°C. They also concluded that for a compound which is relatively volatile and exhibits low solubility in water, such as 1,1,2-trichloroethane, air stripping may be a viable removal technique.

- No actual system performance data on the removal of 1,1,2-trichloroethane by packed column aeration have been found. U.S. EPA (1988) reported a calculated value for Henry's Law constant of 43 atm at 20°C. This value is in agreement with the data of Ashworth et al. (1988) who reported a laboratory measured Henry's Law coefficient of 41.1 atm at 20°C. 1,1,2 Trichloroethane may be amenable to removal by aeration.

- Dobbs and Cohen (1980) developed adsorption isotherms for 1,1,2-trichloroethane on activated carbon Filtrasorb® 300. The isotherm constant for 1,1,2-trichloroethane is 5.81 $mgL^{1/n}/gmg^{1/n}$ while 13 is the value of the isotherm slope $(1/n)$. They reported adsorption capacities of 1.5 mg and 6 mg of 1,1,2-trichloroethane per gram of carbon at influent concentrations of 0.1 mg/L and 1.0 mg/L, respectively. The results indicate that 1,1,2-trichloroethane isotherms were greatly affected by influent concentration and by total organic carbon (TOC) concentration.

- Shuckrow and Pajak (1981, 1982) evaluated the performance of granular activated carbon (GAC) and resin adsorption in removing TOC from contaminated ground water in Michigan. 1,1,2-Trichloroethane concentrations of 0.11 mg/L and 0.07 mg/L were detected in ground water. GAC and resin adsorption, followed by aerobic biological treatment, reportedly removed almost all of the organic priority pollutants, including 1,1,2-trichloroethane, from the ground water to less than the level of detection (0.01 mg/L).

- Beaudet et al. (no date given) conducted a 3-year project to remove priority pollutants from wastewater by GAC. 1,1,2-Trichloroethane was present in concentrations ranging from 0.01 mg/L to 110 mg/L. A pilot-plant adsorption system, containing three columns packed with Filtra-sorb® 300 granular carbon, was operated at 20-, 40- and 60-minute cumulative empty bed contact times (EBCT). 1,1,2-Trichloroethane was effectively removed by column 1 and column 2, while column 3 did not reach breakthrough on this compound prior to the end of the end of the pilot operations.

- Van Dyke et al. (1986) studied and reported the efficiency of a home-use water filter containing pressed carbon block as a filtering media for the removal of a number of organic chemicals, including 1,1,2-trichloroethane. The filtering system consisted of a nonwoven prefilter, a pressed carbon block and a porous polyethylene-fritted core. The water was supplied at a constant pressure of 50 psig. Each run consisted of passing a volume of water equal to 150% of the filter-rated life of 500 gal, and analyzing for the various contaminants. The influent concentration was reported at 7 $\mu g/L$ as the sum for both 1,1,2- and 1,1,1-trichloroethane due to chromatographic overlap. This system removed 1,1,2-trichloroethane to below detection limit (0.1 $\mu g/L$).

- Based upon the similarities in structure and physical/chemical properties between 1,1,2-trichloroethane and 1,1,1-trichloroethane, similar removals would be expected. The following data pertains to 1,1,1-trichloroethane.

  - Reinhard et al. (1986) reported on a full-scale reverse osmosis (RO) treatment plant equipped with cellulose acetate membranes; and two pilot RO plants, equipped with polyamide membranes (PA), were tested to determine this technology capability to remove TOC and a variety of trace organics, including 1,1,1-trichloroethane. The full-scale cellulose acetate RO plant was operated at 460 psi, while the pilot-scale polyamide RO plants were operated at 250 psi. Both systems achieved greater than 90% 1,1,1-trichloroethane reduction from an average influent concentration of approximately 3 $\mu g/L$.

  - One program was developed to evaluate the performance of RO as a method of chemical spill cleanup. The program was designed to test commercially available membranes for their effectiveness in such applications (Whittaker, 1984). Tests carried out with a TFC polyether-polysulphone membrane in the spiral configuration showed more than 99% rejection efficiency for 1,1,1-trichloroethane from an initial concentration of 659 mg/L.

VIII. References

ACGIH. 1986. American Conference of Governmental Industrial Hygienists. TLVs–threshold limit values and biological exposure indices for 1985–1986. Cincinnati, OH: ACGIH.

Archer, W.L. 1979. Chlorocarbons, hydrocarbons (other). In: M. Grayson and D. Eckroth, eds. Kirk-Othmer Encyclopedia of Chemical Technology, Vol. 5., 3rd ed., New York, NY: John Wiley and Sons, Inc. pp. 731–733.

Ashworth, R.A., G.B. Howe, M.E. Mullins and T.N. Rogers. 1988. Air-water partitioning coefficients of organics in dilute aqueous solutions. J. Hazard. Mat. 18:25–37.

Barber, E.D., W.H. Donish and K.R. Mueller. 1981. A procedure for the quantitative measurement of the mutagenicity of volatile liquids in the Ames Salmonella/microsome assay. Mutat. Res. 90:31–48.

Beaudet, B.A., E.M. Kellar, L.J. Bilello and R.J. Turner. n.d. Removal of specific organic contaminants from industrial wastewaters by granular activated carbon adsorption. Cincinnati, Ohio: U.S. EPA.

Callahan, M.A., M.W. Slimak, N.W. Gabel et al. 1979. Water-related environmental fate of 129 priority pollutants. EPA 440/4-79-029b. NTIS PB80-22204381. Vol. II. Washington, DC: U.S. EPA.

CFR. 1985. Code of Federal Regulations. OSHA Occupational Standards. Permissible Exposure Limits. 29 CFR 1910.1000.

Chang, L.W., M.A. Pereira and J.E. Klaunig. 1985. Cytotoxicity of halogenated alkanes in primary cultures of rat hepatocytes from normal, partially hepatectomized, and preneoplastic/neoplastic liver. Toxicol. Appl. Pharmacol. 80:274–283.

Dilling, W.L. 1977. Interphase transfer process. II. Evaporation rates of chloromethanes, ethanes, ethylenes, propanes and propylenes from dilute aqueous solutions. Comparisons with theoretical predictions. Environ. Sci. Technol. 11:405–409.

DiRenzo, A.B., A.J. Gandolfi and I.G. Sipes. 1982. Microsomal bioactivation and covalent binding of aliphatic halides to DNA. Toxicol. Lett. 11:243–252.

Dobbs, R.A. and J.M. Cohen. 1980. Carbon adsorption isotherms for toxic organics. EPA 600/8-80-023. Cincinnati, Ohio: U. S. EPA Office of Research and Development, HERL, Wastewater Research Division.

Duprat, P., L. Delsaut and D. Gradiski. 1976. Pouvoir irritant des principaux solvants chlores aliphatiques sur la peau et les muquenses oculaires du lapin. Eur. J. Toxicol. 9:171–177. (Fre.)

Grant, W.M., ed. 1974. Toxicology of the Eye, 2nd ed. Springfield, IL: Charles C. Thomas. p. 1,034.

Hardie, D.W.F. 1964. Chlorocarbons and chlorohydrocarbons. 1,1,2-Trichloroethane. In: Kirk, R.E. and R.E. Othmer, eds. Encyclopedia of Chemical Technology, Vol. 5, 2nd ed. New York, NY: John Wiley and Sons, Inc. pp. 157–159.

Hawley, G.G. 1981. The Condensed Chemical Dictionary, 10th ed. New York, NY: Van Nostrand Reinhold Co. p. 1,041.

HSDB. 1985. Hazardous Substances Data Bank. 1985. Online: March 3.

IARC. 1979. International Agency for Research on Cancer. IARC monographs on the evaluation of the carcinogenic risk of chemicals to humans. 1,1,2-Trichloroethane, Vol. 20, Lyon, France: IARC, WHO, pp. 533–543.

Ikeda, M. and H. Ohtsuji. 1972. A comparative study of the excretion of Fujiwara reaction-positive substances in urine of humans and rodents given trichloro-or tetrachloro-derivatives of ethane and ethylene. Br. J. Indust. Med. 29:99–104.

Jakobson, I., B. Holmberg and J.E. Wahlberg. 1977. Variations in the blood concentration of 1,1,2-trichloroethane by percutaneous absorption and other routes of administration in the guinea pig. Acta Pharmacol. Toxicol. 41:497–506.

Konemann, H. 1981. Quantitative structure-activity relationships in fish toxicity studies. Part 1: Relationship for 50 industrial pollutants. Toxicology. 19:209–221.

Krill, R.M. and W.D. Sonzogni. 1986. Chemical monitoring of Wisconsin's ground water. J. AWWA. 78:70–75.

Leighton, D.T., Jr. and J.M. Calo. 1981. Distribution coefficients of chlorinated hydrocarbons in dilute air-water systems for groundwater contamination applications. J. Chem. Engin. Data 26: 382–385.

Lundberg, I., M. Ekdahl, T. Kronevi, V. Lidums and S. Lundberg. 1986. Relative hepatotoxicity of some industrial solvents after intraperitoneal injection or inhalation exposure in rats. Environ. Res. 40:411–420.

Mitoma, C., T. Steeger, S.E. Jackson, K.P. Wheeler, J.H. Rogers and H.A. Milman. 1985. Metabolic disposition study of chlorinated hydrocarbons in rats and mice. Drug Chem. Toxicol. 8:183–194.

NCI. 1978. National Cancer Institute. Bioassay of 1,1,2-trichloroethane for possible carcinogenicity. U.S. DHEW Publ. No. (NIH) 78-1324. PHS, National Institutes of Health.

Page, G.W. 1981. Comparison of ground water and surface water for patterns and levels of contamination by toxic substances. Environ. Sci. Technol. 15:1474–1481.

Reinhard, M., N.L. Goodman, P.L. McCarty and D.G. Argo. 1986. Removing trace organics by reverse osmosis using cellulose acetate and polyamide membranes. J. AWWA. 78:163–174.

Sanders, V.M., K.L. White, Jr., G.M. Shopp, Jr. and A.E. Munson. 1985. Humoral and cell-mediated immune status of mice exposed to 1,1,2-trichloroethane. Drug Chem. Toxicol. 8:357–372.

Sato, A. and T. Nakajima. 1979. A structure-activity relationship of some chlorinated hydrocarbons. Arch. Environ. Health. 34:69–75.

Seidenberg, J.M., D.G. Anderson and R.A. Becker. 1986. Validation of an in vivo developmental toxicity screen in the mouse. Teratogen., Carcinogen., and Mutagen. 6:361–374.

Shuckrow, A.J. and A.P. Pajak. 1981. Bench scale assessment of concentration technologies for hazardous aqueous waste treatment EPA-600/9-81-0026, Cincinnati, Ohio. U.S. Environmental Protection Agency.

Shuckrow, A.J. and A.P. Pajak. 1982. Bench scale treatability of contaminated groundwater at the Ott/Story Site–Part 1. J. Hazard. Mat. 7:37–50.

Simmon, V.F., K. Kauhanen and R.G. Tardiff. 1977. Mutagenic activity of chemicals identified in drinking water. In: Scott D. et al., eds. Progress in Genetic Toxicology. Amsterdam, Netherlands: Elsevier/North Holland Biomedical Press.

Smyth, H.F., Jr., C.P. Carpenter, C.S. Weil, U.C. Pozzani, J.A. Striegel and J.S. Wycum. 1969. Range-finding toxicity data: List VII. Am. Ind. Hyg. Assoc. J. 30:470–476.

Tabak, H.H., S.A. Quave, C.I. Mashni and E.F. Barth. 1981. Biodegradability studies with organic priority pollutant compounds. J. Water Pollut. Control Fed. 53:1503–1518.

Takano, T., Y. Miyazaki and Y. Motohashi. 1985. Interaction of trichloroethane isomers with cytochrome P-450 in the perfused rat liver. Fund. Appl. Toxicol. 5:353–360.

Thompson, J.A., B. Ho and S.L. Mastovich. 1984. Reductive metabolism of 1,1,1,2-tetrachloroethane and related chloroethanes by rat liver microsomes. Chem.-Biol. Interact. 51:321–333.

Thompson, J.A., B. Ho and S.L. Mastovich. 1985. Dynamic headspace analysis of volatile metabolites from the reductive dehalogenation of trichloro- and tetrachloroethanes by hepatic microsomes. Anal. Biochem. 145:376–384.

Tu, A.S., T.A. Murray, K.M. Hatch, A. Sivak and H.A. Milman. 1985. In vitro transformation of BALB/c-3T3 cells by chlorinated ethanes and ethylenes. Cancer Lett. 28:85–92.

Tyson, C.A., K. Hawk-Prather, D.L. Story and D.H. Gould. 1983. Correlations of in vitro and in vivo hepatotoxicity for five haloalkanes. Toxicol. Appl. Pharmacol. 70:289–302.

U.S. EPA. 1975. Analysis of carbon and resin extracts. New Orleans area water supply study.

U.S. EPA. 1980. Ambient water quality criteria document for chlorinated ethanes. Prepared by the Office of Health and Environmental Assessment, Environmental Criteria and Assessment Office, Cincinnati, OH for the Office of Water Regulations and Standards, Washington, DC. EPA 440/5-80-029. NTIS PB81-117400.

U.S. EPA. 1983. Hazard profile for 1,1,2-trichloroethane. Prepared by the Office of Health and Environmental Assessment, Environmental Criteria and Assessment Office, Cincinnati, OH for the Office of Solid Waste, Washington, DC.

U.S. EPA. 1985a. Method 502.1 — Volatile halogenated organic compounds in water by purge and trap gas chromatography. Cincinnati, OH: Environmental Monitoring and Support Laboratory, June 1985 (Revised November 1985).

U.S. EPA. 1985b. Method 524.1 — Volatile organic compounds in water by purge and trap gas chromatographic/mass spectrometry. Cincinnati, OH: Environmental Monitoring and Support Laboratory, June 1985 (Revised November 1985).

U.S. EPA. 1986. Guidelines for carcinogen risk assessment. Fed. Reg. 51(185):33992-34003.

U.S. EPA. 1988. Preliminary assessment of the Best Available Technology for the removal of Phase V synthetic organic chemicals from potable water supplies. First Draft. Prepared for U.S. EPA, Science and Technology Branch.

Van Dyke, K., R. Kuennen, J. Stiles, J. Wzeman, and J. O'Neal. 1986. Test stand design and testing for a pressed carbon block water filter. Amer. Lab. 18:118-132.

Van Dyke, R.A. and C.G. Wineman. 1971. Enzymatic dechlorination: dechlorination of chloroethanes and propanes in vitro. Biochem. Pharmacol. 20:463-470.

Verschueren, K. 1983. Handbook of Environmental Data on Organic Chemicals, 2nd. ed. New York, NY: Van Nostrand Reinhold Co. pp. 1128-1129.

Wahlberg, J.E. 1984a. Erythema-inducing effects of solvents following epicutaneous administration to man: studied by laser Doppler flowmetry. Scand. J. Work Environ. Health. 10:159-162.

Wahlberg, J.E. 1984b. Edema-inducing effects of solvents following topical administration. Dermatosen Beruf. Umwelt. 32:91-94.

Weast, R.C. and M. Astle, eds. 1986. CRC Handbook of Chemistry and Physics-A Ready Reference Book of Chemical and Physical Data, 67th ed. Cleveland, OH: CRC Press.

White, K.L., Jr., V.M. Sanders, D.W. Barnes, G.M. Shopp, Jr. and A.E.

Munson. 1985. Toxicology of 1,1,2-trichloroethane in the mouse. Drug Chem. Toxicol. 8:333–355.

Whittaker, H. 1984. Development of a mobile reverse osmosis unit for spill cleanup. Presented at the Hazardous Material Spills Conference, Prevention, Behavior, Control and Cleanup of Spills and Waste Sites. Rockville, MD: Government Institutes, Inc.

Wilson, J.T., C.G. Enfield, W.J. Dunlap, R.L. Cosby, D.A. Foster and L.B. Baskin. 1981. Transport and fate of selected organic pollutants in a sandy soil. J. Environ. Qual. 10:501–506.

Wilson, J.T., J.F. McNabb, D.L. Balkwill and W.C. Ghiorse. 1983. Enumeration and characterization of bacteria indigenous to a shallow water table aquifer. Groundwater 21:134–142.

Windholz, M., S. Budavari, R.F. Blumetti and E.S. Otterbein, eds. 1983. The Merck Index: An Encyclopedia of Chemicals, Drugs, and Biologicals. Rahway, New Jersey: Merck and Co., Inc. p. 1378.

Wright, W.H. and J.M. Schaffer. 1932. Critical anthelmintic tests of chlorinated alkyl hydrocarbons and a correlation between the anthelmintic efficacy, chemical structure and physical properties. Am. J. Hyg. 16:325–428.

Yllner, S. 1971. Metabolism of 1,1,2-trichloroethane-1,2-^{14}C in the mouse. Acta Pharmacol. Toxicol. 30:248–256.

Zoeteman, B.C.J., K. Harmsen, J.B.H.J. Linders, C.F.H. Morra and W. Slooff. 1980. Persistent organic pollutants in river water and ground water of the Netherlands. Chemosphere 9:231–249.

Trichlorofluoromethane

I. General Information and Properties

A. CAS No. 75–69–4

B. Structural Formula

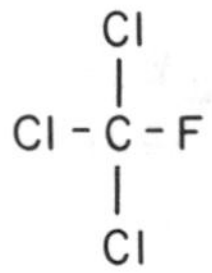

```
      Cl
      |
Cl - C - F
      |
      Cl
```

Trichlorofluoromethane

C. Synonyms

- Trichlorofluoromethane; fluorocarbon 11; Freon-11: F-11; R-11.

D. Uses

- Fluorotrichloromethane (F-11) is used primarily as a plastic foam blowing agent (50–60%), refrigerant (25–30%) and solvent/degreasing agent in the aerospace and electronics industry (5–10%) (WHO, 1986). Miscellaneous uses (1–10%) include heat transfer fluids, fire extinguishers and chemical intermediates. Aerosol propellant use accounted for approximately 80% of production in 1972, but is now minor due to government restrictions (WHO, 1986).

E. Properties (Amoore and Hautala, 1983; Verschueren, 1983; WHO, 1986)

Chemical Formula	CCl_3F
Molecular Weight	137.38
Physical State	Gas at 25°C
Boiling Point (at 1 atm)	23.8°C
Melting Point	−111.0°C
Vapor Pressure (20°C)	687 mm Hg
Specific Gravity (17°C)	1.494

Water Solubility (25°C)	1,100 mg/L
Log Octanol/Water Partition Coefficient	2.53
Taste Threshold	—
Odor Threshold (water)	—
Odor Threshold (air)	5 ppm
Conversion Factor	1 ppm = 5.62 mg/m^3
	1 mg/m^3 = 0.178 ppm

F. Occurrence

- The mean concentration of F-11 in several samples of Lake Ontario water was determined to be 249 ng/L (Kaiser et al., 1983).

- Singh et al. (1979) measured F-11 concentrations in various locations in the Pacific Ocean. The average surface concentration of F-11 was 0.28 ±0.15 ng/L (1.6 ng/m^3). The average concentration at a 300 mg depth was 0.21 (1.2 ng/m^3).

- F-11 was detected in surface snow and rainwater in Alaska (Su and Goldberg, 1976).

- Singh et al. (1979) and Tyson et al. (1978) measured tropospheric levels of fluorocarbons in the northern and southern hemispheres. Mean atmospheric levels of 119–133 ppt (668 ng/m^3) for F-11 were found, with the higher levels in the northern hemisphere.

G. Environmental Fate

- The dominant removal process of F-11 from water is most likely volatilization (WHO, 1986).

- Although volatilization is believed to be the major transport process for F-11, the log octanol/water partition coefficient of F-11 (log K_{ow} = 2.53) indicates that adsorption onto organic particulates may be possible. However, data are inconclusive as to whether F-11 significantly adsorbs onto sediments (U.S. EPA, 1979).

- The hydrolysis rate constant for F-11 is too slow to be measured at 30°C by an unspecified analytical method (DuPont de Nemours and Co., 1980). No specific information was found indicating that F-11 under ambient conditions hydrolyzes rapidly enough for this process to be important in the environmental fate of the compound (U.S. EPA, 1979).

- F-11, at concentrations of 5 and 10 mg/L, was not subject to significant aerobic biodegradation in a lab-scale static-screening flask test method municipal sludge study. The authors attributed this to the extensive rate of volatilization (Tabak et al., 1981).

- Little information is available pertaining to the bioaccumulation of F-11. Dickson and Riley (1976) found F-11 at levels of 0.1–5.0 ppb (0.6–28 $\mu g/m^3$, dry weight basis) in various organs of fish and mollusks. These levels, however, do not necessarily indicate a potential for bioaccumulation. Neely et al. (1974) theorized that bioaccumulation is directly related to the log octanol/water partition coefficient of the compound. The experimentally determined log K_{ow} of 2.53 for F-11 indicates that F-11 is lipophilic and may possibly bioaccumulate in organisms.

II. Pharmacokinetics

A. Absorption

- Quantitative data on the absorption of orally administered F-11 could not be located in the available literature. F-11 would be expected to be absorbed from the gastrointestinal tract on the basis of its physical properties; that is, low molecular weight and relatively high lipid solubility (WHO, 1986).

- The average pulmonary retention of inhaled F-11, measured in three volunteers at apparent steady-state after 1 hour of exposure to 657 mL/m^3 (ppm) (3,690 mg/m^3), was 18.2% (Angerer et al., 1985). The average pulmonary ventilation rate of these volunteers was 9.4 L/minute.

B. Distribution

- Pertinent data regarding the distribution of orally administered F-11 could not be located in the available literature.

- Following 5-minute inhalation exposures of rats to 1–3% (60–170 g/m^3) F-11 in air, the highest tissue levels of F-11 were found in the adrenals, followed by the fat and then the heart (Allen and Hansburys, Ltd., 1971).

- Paulet et al. (1975) administered 2.5% and 5% F-11 by inhalation for 10 minutes to dogs and rabbits (number, sex and strain not specified). In rabbits, there was almost no trace of F-11 in the blood after 2 minutes for the low dose and 10 minutes for the high dose. F-11 appeared rapidly in the bile and urine. Similar results were reported for the dog except that the presence of F-11 in the blood was more prolonged, taking more than 30 minutes for all traces to disappear. F-11 was detectable in cerebrospinal fluid 2 minutes following inhalation.

- Paulet et al. (1975) determined that 30 minutes after a 15-minute inhalation of 0.2% F-11 by dogs (number, sex and strain not specified) only 16% of the inhaled F-11 remained in the body.

C. Metabolism

- Data regarding the metabolism of orally administered F-11 could not be located in the available literature.

- Blake and Mergner (1974) found little evidence of metabolism in an inhalation study in beagles. When the anesthetized beagles breathed 1,040–5,500 ppm (5,800–31,000 mg/m^3) ^{14}C-labeled F-11 from an inhalation bag for 6–23 minutes (total inhaled doses = 410–2,880 mg), essentially all the administered F-11 was recovered unchanged in the expired air.

- *In vitro* studies showed the reductive dechlorination of F-11 to F-21 (fluorodichloromethane) by rat, mouse, rabbit and hamster liver microsomes (Wolf et al., 1978).

D. Excretion

- Data regarding the excretion of F-11 after oral administration could not be located in the available literature.

- Blake and Mergner (1974) reported that virtually all the ^{14}C-labeled F-11 inhaled by anesthetized beagles was recovered unchanged in the expired air. The dogs breathed 1,040–5,500 ppm (5,800–31,000 mg/m^3) for 6–23 minutes (total inhaled doses = 410–2,880 mg; body weights = 9–14 kg). Expired air was collected during and for up to 40 minutes after exposure. Urine was collected for 3 days following exposure; selected dogs were killed 24 hours after exposure for determination of tissue radioactivity. Less than 1% of the administered radioactivity was recovered as ^{14}CO$_2$ and nonvolatile urinary and tissue radioactivity.

- After exposure of three volunteers to F-11 at a mean concentration of 657 mL/m^3 (ppm) (3,690 mg/m^3) for 150–267 minutes, Angerer et al. (1985) found that < 0.01% of the absorbed F-11 was excreted in the urine and that urinary fluoride levels were not elevated. The initial and terminal phases of elimination of F-11 in venous blood were 11 minutes and 1.0 hour and in alveolar air 7 minutes and 1.8 hours, respectively.

III. Health Effects

A. Humans

1. Short-term Exposure

- The only information on oral toxicity is a case report (Haj et al., 1980) of a 38-year-old man who accidentally ingested an unquantified small amount of F-11. He suffered necrosis and multiple perforations of the stomach, requiring corrective surgery. Two days after surgery, he developed moderate jaundice with some signs of hepatocellular injury, namely, elevated serum glutamic oxalacetic transaminase (SGOT) and serum glutamic pyruvic transaminase (SGPT) levels. The authors conjectured that the stomach lesions were caused by the freezing action of F-11. The patient recovered, and a 12-month followup period revealed no abnormalities in liver function or in hematology tests.

- Stewart et al. (1978) exposed groups of 8–10 male and female volunteers ranging in age from 18–46 years old to 0, 250, 500 or 1,000 ppm (0, 1,400, 2,800 and 5,600 mg/m^3) of F-11 for 1–10 hours in a controlled-environment chamber. The F-11 exposures produced no effects on hematological, clinical chemistry and urinalysis values; pulmonary function; EKGs; neurological test values; EEGs; visual-evoked responses; and cognitive performance tests.

- Stewart et al. (1978) also performed repeated-exposure experiments with eight male volunteers in which the same measurements were tested. Exposures were to 0 or 1,000 ppm (5,600 mg/m^3) F-11, 8 hours/day, 5 days/week for approximately 4 weeks. Repeated exposure to F-11 was associated with statistically significant decrements in performance on cognitive tests. No other effects were observed.

- A study by Marier et al. (1973) involved a group of 20 housewives who were given 13 household products containing fluorocarbon propellant (including F-11). The study was divided into three periods: (1) no use or light use of aerosol products for 2 weeks (preexposure period); (2) 4 weeks of heavy use defined as using 21.6 g of propellant/day/subject, which was estimated to be 2–25 times higher than normal use (exposure period); and (3) no use or light use of aerosol products for 2 weeks (postexposure period). All women received physical examinations at the end of the first 2 weeks (preexposure), 6 weeks (exposure) and 8 weeks (postexposure). The only effect noted was an increase within normal limits in lactic dehydrogenase (LDH) during the exposure period. Because there was no accompanying increase in SGOT, Marier et al. (1973) did not consider the increase in LDH to be significant.

2. Long-term Exposure

- Human studies to assess the chronic toxicity of fluorocarbons have not been found in the literature (WHO, 1986).

B. Animals

1. Short-term Exposure

- Kudo et al. (1971) reported LD_{50}s for i.p. injection of F-11 in mice (number and strain not specified) as being 1,743 mg/kg for male mice and 1,871 mg/kg for female mice.

- Slater (1965) investigated the hepatotoxicity of F-11 to female albino rats (number not specified). F-11 was administered in a 1:1 mixture with liquid paraffin as a single gavage dose of 0.5 mL mixture/100 g bw. The dose of F-11 was calculated to be 3,735 mg/kg bw (0.5 mL/100 g bw $\times$ $1/2 \times 1.494$, where 1.494 is the specific gravity of liquid F-11 at 17°C). Controls were not treated with the vehicle. F-11 produced no change in liver NADP and $NADPH_2$ levels measured 1 hour after treatment, nor was there evidence of liver necrosis or change in serum β-glucuronidase activity, determined 3 and 24 hours after treatment. The dose used in this study, 3,735 mg/kg bw, is, thus, a No-Observed-Adverse-Effect Level (NOAEL) for certain liver effects.

- Kudo et al. (1971) administered oral doses of approximately 14, 55 and 220 mg/kg bw/day of F-11 to male and female mice (number and strain not specified) for 1 month. They reported a slight decrease in food consumption and one case of liver vacuoles in the largest dose. There were no abnormalities in urinalysis or hematology for any of the doses.

- In dose-selection testing for a chronic bioassay, groups of five male and five female Osborne-Mendel rats and five male and five female B6C3F$_1$ mice were intubated with 0, 1,000, 1,780, 3,160, 5,620 and 10,000 mg/kg/day of F-11, 5 days/week for 6 weeks, followed by 2 weeks of observation (NCI, 1978). Male rats had a significant depression (26%) of body weight at 1,000 mg/kg/day and at least one death/group at $\geq$ 1,780 mg/kg/day, while female rats had significant body weight depression at 1,780 mg/kg/day and at least one death/group at $\geq$ 3,160 mg/kg/day. Male mice had no body weight depression at $\leq$ 5,620 mg/kg/day and some deaths at $\geq$ 5,620 mg/kg/ day. Female mice had no body weight depression at $\leq$ 3,160 mg/kg/day and some deaths at $\geq$ 3,160 mg/kg/day. No other measurements were done.

2. Dermal/Ocular Effects

- Local chilling or freezing of tissue is a potential problem from liquid F-11 exposure to the skin or the eye (Waritz, 1971).

- No dermal effects were seen on the shaved skin of rabbits sprayed with 40% F-11 in sesame oil daily for 12 days (Scholz, 1962).

- Spray application of F-11 to the skin, tongue, soft palate and auditory canal of rats 1–2 times/day, 5 days/week for 5–6 weeks produced slight irritation of the skin and no other significant effects (Quevauviller et al., 1964; Quevauviller, 1965).

3. Long-term Exposure

- Jenkins et al. (1970) exposed eight male and seven female Sprague-Dawley rats, eight male and seven female Princeton guinea pigs, two male beagle dogs and nine male squirrel monkeys continuously to 0 or 1,000 ppm (5,600 mg/m³) F-11 by inhalation for 90 days. Body weights and hematological values were unaffected. No renal lesions occurred in the dogs, but blood urea nitrogen levels were elevated in the two treated dogs as compared with controls.

- F-11 was administered in corn oil by gavage to groups of 50 male and 50 female Osborne-Mendel rats and 50 male and 50 female B6C3F₁ mice (NCI, 1978). Calculated average dosages, administered 5 days/week for 78 weeks, were 488 and 977 mg/kg/day for male rats, 538 and 1,077 mg/kg/day for female rats and 1,962 and 3,925 mg/kg/day for both male and female mice. Controls consisted of 20 untreated animals/sex/species and 20 vehicle-treated animals/sex/species. After termination of treatment, rats were observed for an additional 28–33 weeks and mice were observed for an additional 13 weeks. A dose-related increase in mortality, relative to vehicle controls, occurred in both male and female rats (p < 0.001, Tarone test). Body weights of treated rats were consistently lower than those of vehicle controls. Signs of illness (hunched appearance and labored respiration) were more frequent in treated groups than in control groups during the first 30 weeks of test. Treatment-related lesions revealed by comprehensive necropsies and histopathological examinations were low incidences of pleuritis and pericarditis, which appeared to be dose-related. Chronic murine pneumonia was observed in 88–100% of the rats in each group, including controls. The increased mortality of treated rats may have been due to a decreased resistance to infection leading to pneumonia. Mortality could not be attributed to changes in body weight, clinical signs or nontumor pathology.

- The only finding in mice was a dose-related increase in mortality in the females as compared with vehicle controls (p = 0.009, Tarone test). The lowest dosage tested, 488 mg/kg/day (5 days/week), was associated with an acceleration of mortality in male rats starting by week 15 of treatment (NCI, 1978).

4. Reproductive Effects

- Pertinent data regarding the reproductive effects of F-11 could not be located in the available literature.

5. Developmental Effects

- Paulet et al. (1974) exposed groups of 20 pregnant Wistar rats and 10 pregnant rabbits to a 20% (200,000 ppm, 1.1 kg/m³) concentration of a 1:9 mixture of F-11 and F-12 in air for 2 hours/day on days 4–16 of gestation (rat) or days 5–20 of gestation (rabbit). Half of the animals in each group were sacrificed at day 20 (rats) or day 30 (rabbits) of gestation, and the rest were allowed to deliver. There were no treatment-related adverse effects on maternal or fetal body weights, number of implantations, resorptions, fetuses, stillbirths, weight of pups at birth and number of pups surviving at 1 and 4 weeks. No anomalies were observed in litters of treated animals.

6. Mutagenicity

- F-11 was not mutagenic to several strains of *Salmonella typhimurium* and *Escherichia coli* K12 when tested in either the presence or absence of a metabolic activating system, or to cultured Chinese hamster ovary cells (Uehleke et al., 1977; Krahn et al., 1982; Longstaff et al., 1984). Negative results were also obtained in a cell transformation assay with BHK21 cells (Longstaff et al., 1984).

7. Carcinogenicity

- When administered by gavage in corn oil to groups of 50 male and 50 female B6C3F₁ mice in a chronic study (NCI, 1978), F-11 produced no evidence of carcinogenicity. Calculated average doses were 0, 1,962 and 3,925 mg/kg/day, 5 days/week for 78 weeks, followed by 13 weeks of observation.

- Gavage administration of F-11 to groups of 50 male and 50 female Sprague-Dawley rats in a chronic study (NCI, 1978) produced no evidence of carcinogenicity, but these results were considered inconclusive by the NCI (1978) because inadequate numbers of rats survived long enough to be at risk from late-developing tumors. Early mortality was largely due to murine pneumonia, which affected 88–100% of the rats.

Calculated average doses administered 5 days/week for 78 weeks (followed by 28–33 weeks of observation), were 0, 488 and 977 mg/kg/day for the males and 0, 538 and 1,077 mg/kg/day for the females.

IV. Quantification of Toxicological Effects

A. One-day Health Advisory

No suitable studies were located in the available literature to calculate a One-day HA. The Ten-day HA of 7 mg/L is recommended as a conservative estimate for a 1-day exposure.

B. Ten-day Health Advisory

Two studies were found that could be used to derive a Ten-day HA: the 1-month oral study (in Japanese) by Kudo et al. (1971), and the 6-week range finding oral gavage study by NCI (1978). The more recent study (NCI, 1978) has been selected as the basis for the Ten-day HA.

The Ten-day HA for a 10-kg child is calculated as follows:

$$\text{Ten-day HA} = \frac{(1,000 \text{ mg/kg/day}) (10 \text{ kg}) (5 \text{ days})}{(1,000) (1 \text{ L/day}) (7 \text{ days})} = 7.1 \text{ mg/L (rounded to } 7,000 \text{ } \mu\text{g/L}$$

where: 5/7 days = dose conversion to represent daily exposure.

The resulting Ten-day HA of 7 mg/L is in substantial agreement with the NAS (1980) 7-day Suggested-No-Adverse-Response Level (SNARL) of 8 mg/L, which was based on the study by Kudo et al. (1971).

C. Longer-term Health Advisory

No suitable studies were located in the available literature to support determination of a Longer-term HA. The Drinking Water Equivalent Level (DWEL) may be adopted as a conservative estimate of the Longer-term HA for a 70-kg adult: 12.2 mg/L, rounded to 10,000 μg/L. This may be modified for a Longer-term HA for a 10-kg child, consuming an average of 1 L/day:

The Longer term HA for a child is calculated as follows:

$$\text{Longer-term HA} = \frac{(0.349 \text{ mg/kg/day}) (10 \text{ kg})}{(1 \text{ L/day})} = 3.49 \text{ mg/L (rounded to } 3,000 \text{ } \mu\text{g/L)}$$

The Longer-term HA for an adult is calculated as follows:

$$\text{Longer-term HA} = \frac{(0.349 \text{ mg/kg/day}) (70 \text{ kg})}{(2 \text{ L/day})} = 12.2 \text{ mg/L (rounded to } 10,000 \text{ } \mu g/L)$$

D. Lifetime Health Advisory

Only one oral study of long-term duration, the NCI (1978) chronic study, was found in the available literature. In this study, F-11 was administered by gavage at 0, 488 and 977 mg/kg/day to male and female rats for 78 weeks. An acceleration of mortality occurred at both F-11 dosage levels in both sexes, relative to vehicle controls. In addition, body weights of treated rats were somewhat lower than those of vehicle controls, overt signs of illness were more frequent in treated groups than in control groups and pleuritis and pericarditis occurred at low incidences in treated groups but not in controls. It should be noted that chronic murine pneumonia occurred in 88–100% of the rats in each group, including controls and, according to NCI (1978), appeared to be a factor in the early mortality. A possible explanation for the preferential acceleration of mortality among treated groups is that F-11 diminished the rats' resistance to pneumonia.

Although the lowest dosage tested in the NCI (1978) study would appear to be a Frank-Effect Level, reconsideration of the available data from U.S. EPA (1982) led to a calculation of the Lifetime HA from the lowest dosage in the chronic oral NCI (1978) study. This dosage of 488 mg/kg/day was designated a LOAEL associated with accelerated mortality in male rats.

From the NCI (1978) study, the Lifetime HA is derived as follows:

Step 1: Determination of the Reference Dose (RfD)

$$\text{RfD} = \frac{(488 \text{ mg/kg/day}) (5 \text{ days})}{(1,000) (7 \text{ days})} = 0.349 \text{ mg/kg/day}$$

where: 5/7 = conversion of dose to represent daily exposure.

Step 2: Determination of the Drinking Water Equivalent Level (DWEL)

$$\text{DWEL} = \frac{(0.349 \text{ mg/kg/day}) (70 \text{ kg})}{(2 \text{ L/day})} = 12.2 \text{ mg/L (rounded to } 10,000 \text{ } \mu g/L)$$

Step 3: Determination of the Lifetime Health Advisory

$$\text{Lifetime HA} = (12.2) (0.2) = 2.44 \text{ mg/L (rounded to } 2,000 \text{ } \mu g/L)$$

E. Evaluation of Carcinogenic Potential

- The chronic gavage study of NCI (1978) was negative for the carcinogenicity of F-11 to B6C3F$_1$ mice and inconclusive for carcinogenicity to Osborne-Mendel rats because of early mortality. No other carcinogenicity studies of F-11 were found in the available literature. IARC has not evaluated the carcinogenic potential of F-11.

- Applying the criteria described in the U.S. EPA's Guidelines for Carcinogen Risk Assessment (U.S. EPA, 1986a), F-11 may be classified in Group D: not classifiable. This category is for agents with inadequate animal evidence of carcinogenicity.

V. Other Criteria, Guidance and Standards

- The NAS (1980) SNARL for 1-day exposure of a 70-kg adult to F-11 is 88 mg/L, based on the study of Slater (1965) and an uncertainty factor of 1,000. NAS (1980) calculated the NOAEL for F-11 to be 2,500 mg/kg bw. For 7-day exposure of a 70-kg adult to F-11, the NAS (1980) SNARL is 8.0 mg/L, based on the study of Kudo et al. (1971), and an uncertainty factor of 1,000.

- The ACGIH (1985) has adopted a ceiling Threshold Limit Value (TLV) of 1,000 ppm (approximately 5,600 mg/m^3) for F-11. The current OSHA standard (PEL) for occupational exposure to F-11 is also 1,000 ppm (Code of Federal Regulations, 1985).

VI. Analytical Methods

- Analysis of F-11 is by a purge-and-trap gas chromatographic procedure used for the determination of volatile organohalides in drinking water (U.S. EPA, 1985a). This method consists of bubbling an inert gas through the sample and trapping volatile compounds on an adsorbent material. The adsorbent material is heated to drive off the compounds onto a gas chromatographic column. The gas chromatograph is temperature programmed to separate the method analytes that are then detected by a halogen-specific detector. The method detection limit has not been determined for this compound. Confirmatory analysis is by mass spectrometry (U.S. EPA, 1985b). The detection limit for confirmation by mass spectrometry is 0.21 μg/L.

VII. Treatment Technologies

- Available data indicate that powdered activated carbon (PAC) and air stripping will remove F-11 from contaminated water.

- The efficiency of PAC in removing volatile organics, including F-11, was investigated at DuPont's Chambers Works Wastewater Treatment Plant (Hutton, 1981). Nuchar SA-15 activated carbon was used at a dosage of 114 mg/L for the treatment of 37 million gallons/day, on the average, of wastewater containing 169 mg/L soluble TOC. PAC was fed upstream of the aeration chamber designed for 8 hours aeration time. The results show 80% removal efficiency of F-11 by this process from a concentration of 15 μg/L.

- Stover (1982) presented the pilot-scale test results from air stripping treatment of contaminated water containing volatile organic compounds. The column, 4 inches in diameter and packed with 25 inches of 1/4-inch glass raschig rings, removed 83% of F-11 from a concentration of 45 μg/L at an air-to-water ratio of 9.3. At an air-to-water ratio of 10.7, the same column removed 85% of F-11 from water with a concentration of 20 μg/L.

- U.S. EPA (1986b) estimated the feasibility of removing F-11 from water by air stripping, employing the engineering design procedures and cost model presented at the 1983 National ASCE Conference on Environmental Engineering. Based on chemical and physical properties and assumed operating conditions, 99% removal efficiency of F-11 was reported for a column 5.5 ft in diameter and packed with 19 ft of 1-inch plastic saddles. The air-to-water ratio required to achieve this degree of removal effectiveness was 6:1. Actual system performance data, however, are necessary to determine the feasibility of using air stripping for the removal of F-11.

- In summary, a number of techniques for the removal of F-11 from water have been examined. While the data are not unequivocal, it appears that PAC adsorption is likely to be a successful treatment technique. The amenability of F-11 to aeration has been clearly established. Selection of individual or combinations of technologies to remove F-11 from contaminated drinking water must be based on a case-by-case technical evaluation, and an assessment of the economics involved.

VIII. References

ACGIH. 1985. American Conference of Governmental Industrial Hygienists. TLVs–Threshold limit values for chemical substances in the work environment, adopted by ACGIH with intended changes for 1985–1986. Cincinnati, OH: ACGIH. p. 32.

Allen and Hansburys, Ltd. 1971. An investigation of possible cardiotoxic effects of the aerosol propellants, arctons 11 and 12, Vol. 1. Unpublished report, courtesy of D. Jack, Managing Director, Allen and Hansburys, Ltd. (Cited in WHO, 1986).

Amoore, J.E. and E. Hautala. 1983. Odor as an aid to chemical safety: odor thresholds compared with threshold limit values and volatilities for 214 industrial chemicals in air and water dilution. J. Appl. Toxicol. 3:272–290.

Angerer, J., B. Schroeder and R. Heinrich. 1985. Exposure to fluorotrichloromethane (R-11). Int. Arch. Occup. Environ. Health. 57(1):67–72.

Blake, D.A. and G.W. Mergner. 1974. Inhalation studies on the biotransformation and elimination of [^{14}C]-trichlorofluoromethane and [^{14}C]-dichlorodifluoromethane in beagles. Toxicol. Appl. Pharmacol. 30(3):396–407.

Code of Federal Regulations. 1985. OSHA Occupational Standards Permissible Exposure Limits. 29 CFR 1910.1000.

Dickson, A.G. and J.P. Riley. 1976. The distribution of short-chain halogenated aliphatic hydrocarbons in some marine organisms. Mar. Pollut. Bull. 7:167–169. (Cited in WHO, 1986.)

DuPont de Nemours and Co. 1980. Freon Products Information Booklet B-2 A98825, 12/80. Wilmington, DE: DuPont de Nemours and Co. (Cited in WHO, 1986.)

Haj, M., Z. Burstein, E. Horn and B. Stamler. 1980. Perforation of the stomach due to trichlorofluoromethane (Freon 11) ingestion. Isr. J. Med. Sci. 16(5):392–394.

Hutton, D.G. 1981. Removal of priority pollutants with a combined powdered activated carbon-activated sludge process. In: Chemistry in Water Reuse, Vol. 2. pp. 403–428.

Jenkins, L.J., Jr., R.A. Jones, R.A. Coon and J. Siegel. 1970. Repeated and continuous exposures of laboratory animals to trichlorofluoromethane. Toxicol. Appl. Pharmacol. 16:133–142.

Kaiser, K.L.E., M.E. Comba and H. Huneault. 1983. Volatile halocarbon contaminants in the Niagara River and in Lake Ontario. J. Great Lakes Res. 9(2):212–223. (CA 99(1):76430e.)

Krahn, D.F., F.C. Barsky and K.T. McCooey. 1982. CRO/HGPRT mutation assay: evaluation of gases and volatile liquids. In: Genotoxic Effects of Airborne Agents. Environ. Sci. Res. 25:91–103.

Kudo, K., S. Toida, S. Matsuura, T. Sasaki and H. Kawamura. 1971. Comparison of Freon-11S and Freon 11. Acute, subacute toxicity and irritation of mucous membrane. Toho Igakkai Zasshi. 18(2):363–367. (Jap.) (Cited in NAS, 1977.) (CA 77(21):135950v.)

Longstaff, E., M. Robinson, C. Bradbrook, J.S. Styles and I.F.H. Purchase. 1984. Genotoxicity and carcinogenicity of fluorocarbons: assessment by short-term *in vitro* tests and chronic exposure in rats. Toxicol. Appl. Pharmacol. 72:15–31.

Marier, G., G.I. McFarland and P. Dussault. 1973. A study of blood fluorocarbon levels following exposure to a variety of household aerosols. Household Pers. Prod. Ind. 10:68, 70, 92, 99. (Cited in WHO, 1986.)

NAS. 1977. National Academy of Sciences. Drinking Water and Health. Washington, DC: NAS. pp. 781–783.

NAS. 1980. National Academy of Sciences. Drinking Water and Health, Vol. 3. Washington, DC: NAS. pp. 166–168.

NCI. 1978. National Cancer Institute. Bioassay of trichlorofluoromethane for possible carcinogenicity. DHEW/PUB/NIH/78-1356. NCI-CG-TR-107. NTIS PB-286187. 93 pp.

Neely, W.B. et al. 1974. Partition coefficient to measure bioconcentration potential of organic chemicals in fish. Environ. Sci. Technol. 8:1113-1115. (Cited in WHO, 1986.)

Paulet, G., S. Desbrousses and E. Vidal. 1974. Absence d'effet teratogene des fluorocarbones chez le rat et le lapin. [Absence of teratogenic effects of fluorocarbons in the rat and rabbit.] Arch. Mal. Prof. Med. Trav. Secur. Soc. 35:658–662. (Fre.)

Paulet, G., J. Lanoe, A. Thos, P. Toulouse and J. Dassonville. 1975. Fate of fluorocarbons in the dog and rabbit after inhalation. Toxicol. Appl. Pharmacol. 34(2):204–213.

Quevauviller, A. 1965. Hygiene et securite des puseurs pour aerosols medicamenteux. Prod. Probl. Pharm. (20(1):14–29.) (Cited in WHO, 1986.)

Quevauviller, A., M. Schrenzel and H. Vu-Ngoc. 1964. Tolerance locale (peau, muqueuses, plaies, brulures) chez l'animal, aux hydrocarbures chlorofluores [Local tolerance in animals to chlorofluorinated hydrocarbons]. Therapie. 19:247–263. (Fre.) (Cited in WHO, 1986.)

Scholz. J. 1962. New toxicologic investigation of Freons used as propellants for aerosols and sprays. Fortschr. Biol. Aerosol Forsch. 1957–1961. Ber. Aerosol-Kongr. 4:420–429. (Cited in Waritz, 1971.)

Singh, H.B., L.J. Salas, H. Shigeishi and E. Scribner. 1979. Atmospheric halocarbons, hydrocarbons, and sulfur hexafluoride: global distributions, sources, and sinks. Science. 203:899–903.

Slater, T.F. 1965. A note on the relative toxic activities of tetrachloromethane and trichloro-fluoro-methane on the rat. Biochem. Pharmacol. 14:178–181.

Stewart, R.D., P.E. Newton, E.D. Baretta, A.A. Herrmann, H.V. Forster and R.J. Soto. 1978. Physiological response to aerosol propellants. Environ. Health Perspect. 26:275–285.

Stover, E.L. 1982. Removal of volatile organics from contaminated ground water. In: Nielson, D. ed. Proc. National Symp. Aquifer Restoration and Groundwater Monitoring, 2nd ed. pp. 77–84.

Su, C.-W. and E.D. Goldberg. 1976. Environmental concentrations and fluxes of some halocarbons. In: Windom, H.L. and R.A. Due, eds. Marine Pollutant Transfer. Lexington, MA: Lexington Books, DC Heath and Company. pp. 353–374.

Tabak, H.H., S.A. Quave, C.I. Mashni and E.F. Barth. 1981. Biodegradability studies with organic priority pollutant compounds. J. Water Pollut. Control Fed. 53:1503–1518.

Tyson, B.J. et al. 1978. Interhemispheric gradients of CF_2CL_2, $CFCl_3$, CCl_4 and H_2O. Geophys. Res. Lett. 5:535–538. (Cited in WHO, 1986.)

Uehleke, H., T. Werner, H. Greim and M. Kraemer. 1977. Metabolic activation of haloalkanes and tests *in vitro* for mutagenicity. Xenobiotica. 7(7):393–400.

U.S. EPA. 1979. U.S. Environmental Protection Agency. Water-related environmental fate of 129 priority pollutants. Volume II. Halogenated aliphatic compounds, halogenated ethers, monocyclic aromatics, phthalate esters, polycyclic aromatic hydrocarbons, nitrosamines and miscellaneous compounds. Washington, DC: Office of Water. EPA 440/4–79–029b.

U.S. EPA. 1982. U.S. Environmental Protection Agency. Errata: halomethanes. Ambient water quality criterion for the protection of human health. Prepared by the Office of Health and Environmental Assessment, Environmental Criteria and Assessment Office, Cincinnati, OH, for the Office of Water Regulations and Standards, Washington, DC.

U.S. EPA. 1985a. U.S. Environmental Protection Agency. Method 502.1 – Volatile halogenated organic compounds in water by purge and trap gas chromatography. Cincinnati, OH: Environmental Monitoring and Support Laboratory. June (Revised November) 1985.

U.S. EPA. 1985b. U.S. Environmental Protection Agency. Method 524.1 – Volatile organic compounds in water by purge and trap gas chromatography/mass spectrometry. Cincinnati, OH: Environmental Monitoring and Support Laboratory. June (Revised November) 1985.

U.S. EPA. 1986a. U.S. Environmental Protection Agency. Guidelines for Carcinogen Risk Assessment. *Fed. Reg.* 51(185):33992–34003.

U.S. EPA. 1986b. U.S. Environmental Protection Agency. Economic evaluation of Freon-11 removal from water by packed column air stripping. Prepared by the Office of Water for Health Advisory Treatment Summaries.

Verschueren, K. 1983. Handbook of Environmental Data on Organic Chemicals, 2nd ed. New York, NY: Van Nostrand Reinhold Co.

Waritz, R.A. 1971. Toxicology of Some Commercial Fluorocarbons. Aerospace Med. Res. Lab., Wright-Patterson Air Force Base, Dayton, OH. NTIS AD751429. (Cited in WHO, 1986.)

Wolf, C.R., L.J. King and D.V. Parke. 1978. The anaerobic dechlorination of trichlorofluoromethane by rat liver preparations *in vitro*. Chem. Biol. Interact. 21(2–3):277-288.

WHO. 1986. World Health Organization. Environmental Health Criteria: chlorofluorocarbons developmental draft document. Prepared by the Environmental Criteria and Assessment Office, U.S. EPA, Cincinnati, OH, for the World Health Organization.

o-Chlorotoluene

I. General Information and Properties

A. CAS No. 95–49–8

B. Structural Formula

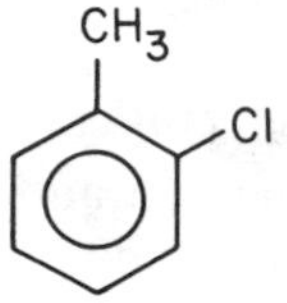

o-Chlorotoluene

C. Synonyms

- 2-Chloro-1-methyl benzene, 2-chlorotoluene, 1-methyl-2-chloroben-zene, 2-methylchlorobenzene, o-tolyl chloride (U.S. EPA, 1985a; NIOSH, 1986; Chemline, 1988).

D. Uses

- o-Chlorotoluene is used as a solvent and as a chemical intermediate in the manufacture of pesticides, dyestuffs, pharmaceuticals and peroxides (Gelfand, 1979).

E. Properties (Gelfand, 1979; Hawley, 1981; Valvani et al., 1981; Amoore and Hautala, 1983; Verschueren, 1983)

Chemical Formula	C_7H_7Cl
Molecular Weight	126.59
Physical State (at 25°C)	Colorless liquid
Boiling Point	159.2°C
Melting Point	-35.6°C
Density	—
Vapor Pressure (20°C)	2.7 mm Hg
Specific Gravity (25°C)	1.0776 g/cm³

37

Water Solubility (20°C)	140 mg/L
Log Octanol/Water Partition Coefficient	3.42
Taste Threshold	—
Odor Threshold (water)	0.0069 mg/L
Odor Threshold (air)	0.32 ppm
	1.7 mg/m³
Conversion Factor	1 ppm = 5.17 mg/m³
	1 mg/m³ = 0.193 ppm

F. Occurrence

- Trace amounts of chlorotoluene (unspecified isomers) were found in 2/15 water samples collected in Buffalo and Niagara Falls, NY (Pellizzari et al., 1979). Elder et al. (1981) found the higher chlorotoluenes, but no monochlorotoluenes, in water and sediment samples collected at sites adjacent to hazardous waste disposal areas in Niagara Falls, NY. Chlorotoluene was also found in the Delaware River in the winter at a level of 3 μg/L (Verschueren, 1983).

- U.S. EPA (1985a) calculated a bioconcentration factor (BCF) of 230 for o-chlorotoluene in aquatic organisms, based on its K_{ow}, indicating that it would be "moderately bioconcentrated." No monitoring data for o-chlorotoluene in possible human food sources were available.

- In Newark, Elizabeth and Camden, NJ, respectively, Harkov et al. (1983) found o-chlorotoluene in 76, 49 and 89% of the air samples collected at mean ambient concentrations of 0.02, 0.02 and 0.01 ppb (0.10, 0.10 and 0.05 μg/m³), respectively. Pellizzari et al. (1979) found chlorotoluene (unspecified isomer) in 80% of the air samples collected in the Love Canal area of Niagara Falls, NY, at concentrations of trace–274 ng/m³. In the same area, Van Tassel et al. (1981) found chlorotoluene isomers in 50% of the air samples at concentrations of < 10 to 1,642 μg/m³. Concentrations of o- and p-chlorotoluene in three other unspecified locations of New York State were 4.6 to 5.8 μg/m³ (Van Tassel et al., 1981). Chlorotoluene isomers were found in 1/11 air samples collected in Baton Rouge, LA, at a concentration of 35 ng/m³ (Pellizzari et al., 1979).

G. Environmental Fate

- The Henry's Law constant calculated for o-chlorotoluene is 0.042 atm-m³ M⁻¹. This value suggests that volatilization may be rapid from all types of surface waters (U.S. EPA, 1985a).

- No data concerning hydrolysis were available, but because chlorine is a ring substituent, hydrolysis is not expected to be an important process in determining environmental fate of o-chlorotoluene (U.S. EPA, 1985a).

- Data concerning biodegradability of o-chlorotoluene are somewhat contradictory. Gibson et al. (1968) found that it was degraded by *Pseudomonas putida* that had been grown in media with toluene as the sole carbon source. On the other hand, Kawasaki (1980) and Sasaki (1978) classified it as degradation resistant using a standardized biodegradation test that was developed for the Japanese Ministry of International Trade and Industry; however, this test often underestimates biodegradability of a compound and, therefore, negative results could be misleading (U.S. EPA, 1985a).

II. Pharmacokinetics

A. Absorption

- Forty-eight hours after 320 mg/kg of uniformly ring-labeled ^{14}C-o-chlorotoluene was administered by gavage to rats, an average of 11% of the administered label was recovered in expired air, 81% in the urine and 4% in the feces, indicating > 90% absorption from the gastrointestinal (GI) tract (Wold and Emmerson, 1974).

- Four days after 1 mg/kg of uniformly ring-labeled ^{14}C-o-chlorotoluene was administered in corn oil by gavage to four female and four male Sprague-Dawley rats, 1.4 and 4.4%, respectively, of the administered label was recovered in the expired air, 85 and 92.5%, respectively, in the urine and 5.3 and 7.6%, respectively, in the feces, indicating > 90% absorption from the GI tract (Quistad et al., 1983). The observation of peak plasma levels of radioactivity at approximately 2 hours posttreatment suggested to the investigators that absorption was rapid.

B. Distribution

- Quistad et al. (1983) measured the levels of radioactivity in tissues of four male and four female rats 4 days after 1 mg/kg single gavage doses in corn oil of uniformly ring-labeled ^{14}C-o-chlorotoluene. Blood levels of radioactivity dropped to < 0.05 ppm (equivalent as o-chlorotoluene) in 10 hours, and about 0 ppm in 24 hours after treatment with 1 mg/kg. In females, 4 days after treatment, skin levels of radioactivity were about 1 ppb o-chlorotoluene and levels in all other tissues measured were < 1 ppb. In results reported for only one male, 4 days after treatment, skin levels of radioactivity were 11 ppb; lung, kidney and pericardial fat levels were 5 to 6 ppb; and all other tissue levels were < 1 ppb. Two additional female rats were treated with 91 or 102 mg/kg of ^{14}C-o-chlorotoluene. Four days after treatment, skin levels of radioactivity were 258 ppb, kidney levels were 168 ppb, liver and lung levels were about 83 ppb, the carcass remains contained 235 ppb, and other organ and tissue levels were < 45 ppb. For all doses, < 1% of the radioactivity remained in the rats after 4 days (all data are means from the two rats).

C. Metabolism

- Quistad et al. (1983) administered single gavage doses of 1, 91 and 102 mg/kg uniformly ring-labeled ^{14}C-o-chlorotoluene to young adult male and female rats and collected urine, feces and expired air for 24 hours. The authors noted that the metabolism of o-chlorotoluene was unaffected by sex or the magnitude of the dose. At the 1 mg/kg dose, the major urinary metabolites were the β-glucuronide of 2-chlorobenzyl alcohol, a mercapturic acid and a glycine conjugate of 2-chlorobenzoic acid at 35 to 42%, 21 to 28% and about 20 to 23% of the total radioactivity in the urine, respectively. Fecal metabolites included the mercapturic acid derivative, 2-chlorohippurate, and the β-glucuronide conjugate of 2-chlorobenzyl alcohol, which were identified in the urine at 22 to 33%, 9 to 18% and 5 to 24% of fecal radioactivity, respectively. Of the radioactivity recovered in expired air < 1% was as carbon dioxide. Unmetabolized ^{14}C-o-chlorotoluene was present as ≥ 84% of the radioactivity in expired air, < 1% of the radioactivity in the urine and ≤ 10% of the radioactivity recovered in the feces, collected in 24 hours. The investigators concluded that the primary metabolic pathway involved oxidation at the methyl group to form 2-chlorobenzyl alcohol or 2-chlorobenzoic acid, which subsequently underwent conjugation with glucuronic acid or amino acids, or with sulfur-containing moieties, to form a mercapturic acid. The recovery of very little radiolabeled carbon dioxide indicated the stability of the benzene ring; the recovery of no radiolabeled hippuric acid indicated the stability of the chloride moiety at the ortho position.

- An arene oxide intermediate may be found in the primary pathway of o-chlorotoluene oxidation, and arene oxides have been associated with tissue necrosis and carcinogenicity (U.S. EPA, 1981).

D. Excretion

- In rats treated by gavage with 1, 91 or 102 mg/kg of uniformly ring-labeled ^{14}C-o-chlorotoluene, about 85 to 92%, 5 to 8% and 1 to 4% of the administered radioactivity were recovered 4 days after treatment in the urine, feces and expired air, respectively (Quistad et al., 1983). Although Quistad et al. (1983) did not report half-lives for o-chlorotoluene, radioactivity was almost completely cleared from the blood of rats within 24 hours of a single gavage treatment with ^{14}C-o-chlorotoluene.

III. Health Effects

A. Humans

- Data regarding the effect of either short-term or long-term human exposure to o-chlorotoluene could not be located in the available literature.

B. Animals

1. Short-term Exposure

- Ely (1971) reported an oral LD_{50} in rats > 1,600 mg/kg and that 100 mg/kg of o-chlorotoluene resulted in "moderate to marked weakness and vasodilation" in rats. No adverse effects were reported in rats treated with 50 mg/kg. However, the experimental protocol and detailed description of effects were not given.

- Pis'ko et al. (1981) reported an LD_{50} in mice, rats and guinea pigs of 4,400, 5,700 and 3,000 mg/kg, respectively. Acute poisoning was manifested as an unsteady gait, damp coat and disheveled fur. Lethal doses caused convulsive jerking and rapid, superficial breathing.

- Rats exposed to 4,000 ppm (20,710 mg/m³) for 6 hours lost coordination after 1.5 hours of treatment, after 1.75 hours of treatment they were prostrate and after 2.0 hours of treatment they developed tremors. Marked vasodilation was also observed (exposure duration not specified). The rats survived and gained weight 14 days after treatment (Ely, 1971).

- Huntingdon Research Center (1982) performed short-term inhalation toxicity studies on 100 adult Sprague-Dawley rats (14 days) and 30 adult New Zealand rabbits (23 days) at doses of about 4,000, 8,000, 12,000 or 16,000 mg/m³ for 6 hours/day exposure to o-chlorotoluene. Each exposure group consisted of 10 male and 10 female rats, and 6 female rabbits.

Three rats died from acute toxicity at $\geq 12,000$ mg/m^3. Dose-related salivation, lachrymation, central nervous system (CNS) depression and ataxia occurred in rats at $\geq 8,000$ mg/m^3. Slight signs of irritation and CNS depression occurred at 4,000 mg/m^3 in rats. Dose-related alopecia and brown staining of fur also occurred in rats. Body weight gain suppression and reduced food consumption were noted in all dose groups in male rats and in rabbits. In male rats, hemoglobin levels were elevated at 4,000 and 16,000 mg/m^3, and packed cell volume was elevated at 16,000 mg/m^3. In rats treated with 4,000 or 16,000 mg/m^3, there was an increase in protein, albumin, serum alkaline phosphatase, serum glutamic pyruvic transaminase and cholesterol in both sexes; sodium and chloride levels were lower for females at 16,000 mg/m^3. Urine volume was significantly higher at 4,000 and 16,000 mg/m^3 (except in low-dose females), and urine pH was significantly lowered at these doses. The liver and kidney weights were significantly increased in rats for all dose groups, except for female rats at 4,000 mg/m^3. Spleen weights of males at $\geq 8,000$ mg/m^3 and females at $\geq 12,000$ mg/m^3 were significantly decreased. The 4,000 mg/m^3 dose has been identified as the Lowest-Observed-Adverse-Effect Level (LOAEL) from this study.

- In a 2-month experiment (Pis'ko et al., 1981), white rats were given 114 or 570 mg/kg of o-chlorotoluene by daily intragastric administration in oil. Unspecified dose-dependent effects on hemopoiesis, liver, kidneys, CNS and immune response were observed.

2. Dermal/Ocular Effects

- o-Chlorotoluene was a moderate irritant when applied to intact or abraded rabbit skin in a 24-hour patch test (Hazelton Laboratories, Inc., 1966).

- One drop of undiluted o-chlorotoluene applied to the eye of a rabbit resulted in a delayed moderate erythema of the conjunctiva and opacity of the anterior portion of the cornea (Ely, 1971).

3. Long-term Exposure

- Three groups of four adult beagle dogs (male and female) were given daily oral doses of 5, 20 or 80 mg/kg o-chlorotoluene in gelatin capsules, containing an aqueous emulsion of o-chlorotoluene in 5% acacia (Gibson et al., 1974b). The control group was given the vehicle only. The duration of the study was 97 days. Dogs were observed daily; hematological, clinical biochemistry and urinalysis profiles were measured at 1, 2 and 4 weeks, and 2 and 3 months. Body weights were measured weekly and the dose adjusted whenever a change of 10% occurred. Organ weights were measured and tissues were examined histopathologically at the end of the 97-day exposure. The only statistically significant change observed in any tested parameters for any treatment group was an increase in platelet count in females treated with 80 mg/kg. Since the platelet count returned to normal after 14 weeks, the authors did not consider the increased platelet count to be treatment related. A NOAEL of 80 mg/kg is indicated for this species from this study.

- Weanling Harlan rats (125 g each) were divided into four groups, each containing 20 animals/sex (Gibson et al., 1974a). For 103 to 104 days, animals were administered by gavage 20, 80 or 320 mg/kg of o-chlorotoluene in an aqueous solution containing 5% acacia as the emulsifying agent. The control group was given the vehicle only. No behavioral changes were observed in any dose groups. Statistically significant decreases in mean body weight gain and an increase in adrenal weight occurred in males fed with 80 and 320 mg/kg o-chlorotoluene. Heart and testes weight were also increased in the male 320 mg/kg dose group. Blood urea nitrogen was increased in the group receiving 80 mg/kg, and an increase in white blood cell count and a decrease in the prothrombin time were observed in male rats treated with 320 mg/kg. Histopathological examination did not reveal any differences between control and experimental groups. A NOAEL of 20 mg/kg has been identified from this study.

4. Reproductive Effects

- Data regarding the effects of reproductive toxicity of o-chlorotoluene could not be located in the available literature.

5. Developmental Effects

- Huntingdon Research Center (1983a) studied the effect of o-chlorotoluene vapor on pregnant rats. Female CrL:COBS CD(SD)Br rats were exposed to o-chlorotoluene vapors for 6 hours/day at concentrations of 0, 1,000, 3,000 or 9,000 mg/m^3 (25 animals/group) during gestation days 6 through 19. On day 20 the animals were killed, litter values determined and fetuses examined for visceral and skeletal abnormalities. Among parent animals exposed to 9,000 mg/m^3, ataxia, lachrymation, salivation, increased water consumption, decreased body weight gain and reduced food consumption were observed. Slight ataxia, reduction in body weight gain and increased water consumption were observed among animals inhaling 3,000 mg/m^3. No effects were observed at 1,000 mg/m^3. Litter size, instances of total resorption and pre- and postimplantation loss were not affected by treatment. At 9,000 mg/m^3, litter and mean fetal weight were significantly reduced and there was an increased incidence of malformed fetuses and fetuses with skeletal anomalies and sternebral variants. Although at 1,000 mg/m³, one fetus was observed to have malformations that were similar to those observed in some fetuses at 9,000 mg/m³, the authors stated that at 1,000 and 3,000 mg/m³, there was no conclusive evidence of treatment-related effects on fetuses.

- Huntingdon Research Center (1983b) studied the effect of o-chlorotoluene vapor inhalation on pregnancy in New Zealand White rabbits. Sixteen rabbits/group were exposed for 6 hours/day to 0, 1,500, 4,000 or 10,000 mg/m³ of o-chlorotoluene vapors during gestation days 6 through 28. On day 29 the animals were killed, litter values determined and fetuses examined for visceral and skeletal abnormalities. At 4,000 and 10,000 mg/m³, there were lachrymation, salivation and ptosis in parent animals. No effects were observed at 1,500 mg/m³. There were no significant or dose-related effects on litter size, pre- and postimplantation losses, mean fetal weight or incidences of malformation in any dose group. A maternal NOAEL of 1,500 mg/m³ and a fetal NOAEL of 10,000 mg/m³ have been identified from this study.

6. Mutagenicity

- Data regarding the mutagenicity of o-chlorotoluene could not be located in the available literature.

7. Carcinogenicity

- Data regarding the carcinogenic effects of o-chlorotoluene could not be located in the available literature. The chemical has not been scheduled for testing by the NTP (1987).

IV. Quantification of Toxicological Effects

A. One-day Health Advisory

Data are insufficient for calculation of a One-day HA for a 10-kg child. The Ten-day HA of 2 mg/L represents a conservative estimate for one-day exposure.

B. Ten-day Health Advisory

The study by Gibson et al. (1974a) has been selected to serve as the basis for the Ten-day HA for a 10-kg child because an oral exposure route was used and the duration of the exposure period (103 to 104 days) is adequate. A NOAEL of 20 mg/kg has been identified from this study. The study by Pis'ko et al.

(1981), although using an oral route of exposure in rats, was not selected because the doses used were higher and a NOAEL was not identified. The LOAEL of 114 mg/kg obtained in Pis'ko et al. (1981), however, is similar to the LOAEL of 80 mg/kg obtained by Gibson et al. (1974a). The Ten-day HA for the 10-kg child is calculated as follows:

$$\text{Ten-day HA} = \frac{(20 \text{ mg/kg/day}) (10 \text{ kg})}{(100) (1 \text{ L/day})} = 2 \text{ mg/L}$$

C. Longer-term Health Advisory

The study by Gibson et al. (1974a) has been selected to serve as the basis for the Longer-term HA because an oral exposure route was used and rats were dosed for 103 to 104 days, an appropriate longer-term exposure period. The Longer-term HA for the 10-kg child is calculated as follows:

$$\text{Longer-term HA} = \frac{(20 \text{ mg/kg/day}) (10 \text{ kg})}{(100) (1 \text{ L/day})} = 2 \text{ mg/L } (2{,}000 \text{ } \mu g/L)$$

The Longer-term HA for the 70-kg adult is calculated as follows:

$$\text{Longer-term HA} = \frac{(20 \text{ mg/kg/day}) (70 \text{ kg})}{(100) (2 \text{ L/day})} = 7 \text{ mg/L } (7{,}000 \text{ } \mu g/L)$$

D. Lifetime Health Advisory

A DWEL may be derived from the Gibson et al. (1974a) subchronic oral exposure study by applying an additional uncertainty factor to account for a less than lifetime exposure.

Step 1: Determination of the Reference Dose (RfD)

$$\text{RfD} = \frac{20 \text{ mg/kg/day}}{1{,}000} = 0.02 \text{ mg/kg/day}$$

Step 2: Determination of the Drinking Water Equivalent Level (DWEL)

$$\text{DWEL} = \frac{(0.02 \text{ mg/kg/day}) (70 \text{ kg})}{2 \text{ L/day}} = 0.7 \text{ mg/kg/day } (700 \text{ } \mu g/L)$$

Step 3: Determination of the Lifetime Health Advisory

$$\text{Lifetime HA} = (0.7 \text{ mg/L}) (20\%) = 0.14 \text{ mg/L(rounded to 0.1 mg/L or 100 } \mu g/L)$$

E. Evaluation of Carcinogenic Potential

- Data were not located regarding the carcinogenicity of o-chlorotoluene. This chemical has not been scheduled for carcinogenicity testing by the NTP (1987).

- IARC has not evaluated the carcinogenic potential of o-chlorotoluene. Applying the criteria described in the U.S. EPA's Guidelines for Carcinogen Risk Assessment (U.S. EPA, 1986a), o-chlorotoluene may be classified in Group D: not classified. This category signifies that available evidence is insufficient to evaluate the agent's carcinogenic potential or no data are available.

V. Other Criteria, Guidance and Standards

- ACGIH (1985, 1980) has recommended a Threshold Limit Value (TLV) of 50 ppm (≈ 250 mg/m^3) for o-chlorotoluene based on analogy to the toxicity of other chlorinated benzenes. Pis'ko et al. (1981) recommended maximum permissible concentrations of 0.04 mg/L of both o- and p-chlorotoluene in water reservoirs.

VI. Analytical Methods

- Analysis of o-chlorotoluene is by a purge-and-trap gas chromatographic procedure used for the determination of volatile aromatic and unsaturated organic compounds in water (U.S. EPA, 1985b). This method calls for the bubbling of an inert gas through the sample and trapping volatile compounds on an adsorbent material. The adsorbent material is heated to drive off compounds onto a gas chromatographic column. The gas chromatograph is temperature programmed to separate the method analytes, which are then detected by the photoionization detector. This method is applicable to the measurement of o-chlorotoluene over a concentration range of 0.08 to 1,500 μg/L. Confirmatory analysis is by mass spectrometry (U.S. EPA, 1985c). The detection limit for confirmation by mass spectrometry has not been determined.

VII. Treatment Technologies

- U.S. EPA (1986b) estimated the feasibility of removing o-chlorotoluene from water by packed column aeration, employing the engineering design procedure and cost model presented at the 1983 National ASCE Conference on Environmental Engineering. Based on chemical and physical properties and assumed operating conditions, a 90% removal efficiency of o-chlorotoluene was reported for a column with a diameter of 5.5 feet and packed with 11 feet of 1-inch plastic saddles. The air-to-water ratio required to achieve this degree of removal effectiveness is approximately 1:10.

- Leighton and Calo (1981) reported experimental measurements of the distribution coefficients for 21 chlorinated hydrocarbons, including o-chlorotoluene, in a dilute air-water system. (The distribution coefficient is the ratio of the volume of a compound in air to the volume of the compound in water after purging.) They determined the distribution coefficient for o-chlorotoluene to be approximately 197.2 at 25°C. They also concluded that for a compound which is relatively volatile and exhibits low solubility in water, such as o-chlorotoluene, air stripping may be a viable removal technique.

- Data were not found for the removal of o-chlorotoluene from drinking water by activated carbon adsorption. However, evaluation of physical/chemical properties of o-chlorotoluene indicate that it may be amenable to removal by activated carbon adsorption due to its low solubility.

VIII. References

ACGIH. 1980. American Conference of Governmental Industrial Hygienists. Documentation of the threshold limit values, 4th ed. (includes supplemental documentation, 1981, 1982, 1983). Cincinnati, OH: ACGIH. pp. 95–96.

ACGIH. 1985. American Conference of Governmental Industrial Hygienists. TLVs — Threshold limit values for chemical substances in the work environment adopted by ACGIH with intended changes for 1985–1986. Cincinnati, OH: ACGIH. p. 13.

Amoore, J.E. and E. Hautala. 1983. Odor as an aid to chemical safety: odor thresholds compared with threshold limit values and volatilities for 214 industrial chemicals in air and water dilution. J. Appl. Toxicol. 3(6)272–290.

Chemline. 1988. Online. Bethesda, MD: National Library of Medicine.

Elder, V.A., B.L. Proctor and R.A. Hites. 1981. Organic compounds found near dump sites in Niagara Falls, New York. Environ. Sci. Technol. 15(10)1237–1243.

Ely, T.S. 1971. Personal communication to TLV Committee from Lab. Ind. Med., Eastman Kodak Co., Rochester, NY. November 3. (Cited in ACGIH, 1980.)

Gelfand, S. 1979. Chlorocarbons, hydrocarbons (toluene). In: Kirk, R.E., and D.F. Othmer, eds. Encyclopedia of Chemical Technology, Vol. 5, 3rd ed. New York, NY: John Wiley and Sons. pp. 819–827.

Gibson, D.T., J.R. Koch, C.L. Schuld and R.E. Kallio. 1968. Oxidative degradation of aromatic hydrocarbons by microorganisms. II. Metabolism of halogenated aromatic hydrocarbons. Biochemistry. 7(11)3795–3802.

Gibson, W.R., F.O. Gossett, G.R. Koenig and F. Marroquin. 1974a. The toxicity of daily oral doses of o-chlorotoluene in the rat. Prepared by Toxicology Division, Lilly Research Laboratories. Submitted to Test Rules Development Branch, Office of Toxic Substances, U.S. EPA, Washington, DC.

Gibson, W.R., F.O. Gossett, G.R. Koenig and F. Marroquin. 1974b. The toxicity of daily oral doses of o-chlorotoluene in the dog. Prepared by Toxicology Division, Lilly Research Laboratories. Submitted to Test Rules Development Branch, Office of Toxic Substances, U.S. EPA, Washington, DC.

Harkov, R., B. Kebbekus, J.W. Bozzelli and P.J. Lioy. 1983. Measurement of selected volatile organic compounds at three locations in New Jersey during the summer season. J. Air Pollut. Control Assoc. 33(12):1177–1183.

Hawley, G.G. 1981. The Condensed Chemical Dictionary, 10th ed. New York, NY: Van Nostrand Reinhold Co. p. 243.

Hazelton Laboratories, Inc. 1966. Acute inhalation exposure–rats, mice, guinea pigs; primary skin irritation–rabbits; acute eye irritation–rabbits. Project Nos. 157-147 and 157-148. (Cited in ACGIH, 1980.)

Huntingdon Research Center. 1982. 2-Chlorotoluene: a preliminary inhalation study in the rat and rabbit. Submitted to Test Rules Development Branch, Office of Toxic Substances, U.S. EPA, Washington, DC.

Huntingdon Research Center. 1983a. Effect of 2-chlorotoluene vapors on pregnancy of the rat. Submitted to Test Rules Development Branch, Office of Toxic Substances, U.S. EPA, Washington, DC.

Huntingdon Research Center. 1983b. Effect of 2-chlorotoluene vapors on pregnancy of the New Zealand White rabbit. Submitted to Test Rules Development Branch, Office of Toxic Substances, U.S. EPA, Washington, DC.

Kawasaki, M. 1980. Experiences with the test scheme under the chemical control law of Japan: an approach to structure-activity correlations. Ecotoxicol. Environ. Safety. 4:444–454.

Leighton, D.T., Jr. and J.M. Calo. 1981. Distribution coefficients of chlorinated hydrocarbons in dilute air-water systems for groundwater contamination application. J. Chem. Eng. Data. 26(4):382–385.

NIOSH. 1986. National Institute for Occupational Safety and Health. RTECS (Registry of Toxic Effects of Chemical Substances). March 1986: Online.

NTP. 1987. National Toxicology Program. Toxicology Research and Testing Program: management status report. Research Triangle Park, NC: NTP. April 15.

Pellizzari, E.D., M.D. Erickson and R.A. Zweidinger. 1979. Formulation of preliminary assessment of halogenated organic compounds in man and environmental media. Research Triangle Park, NC: U.S. EPA. EPA 560/13-79-006.

Pis'ko, G.T., T.V. Tolstopyatova, T.V. Belyanina et al. 1981. Study of maximum permissible concentrations of o- and p-chlorotoluenes in bodies of water. Gig. Sanit. (8):67–68.

Quistad, G.B., K.M. Mulholland and G.C. Jamieson. 1983. 2-Chlorotoluene metabolism by rats. J. Agric. Food Chem. 31(6):1158–1162.

Sasaki, S. 1978. The scientific aspects of the chemical substances control law in Japan. In: Hutzinger, O., L.H. Van Letyoeld and B.C.J. Zoetman, eds. Aquatic Pollutants: Transformation and Biological Effects. Elmsford, NY: Pergamon Press. pp. 283–298.

U.S. EPA. 1981. U.S. Environmental Protection Agency. Eighth report of the Interagency Testing Committee to the Administration; receipt of report and request for comments regarding priority list of chemicals. *Fed. Reg.*46(99):28138–28144.

U.S. EPA. 1985a. U.S. Environmental Protection Agency. Health and environmental effects profile for chlorotoluenes (o-, m-, p-). Prepared by the Office of Health and Environmental Assessment, Environmental Criteria and Assessment Office, Cincinnati, OH for the Office of Solid Waste, Washington, DC.

U.S. EPA. 1985b. U.S. Environmental Protection Agency. U.S. EPA Method 503.1–Volatile aromatic and unsaturated organic compounds in water by purge and trap gas chromatography. Cincinnati, OH: Environmental Monitoring and Support Laboratory. June 1985 (Revised November 1985.)

U.S. EPA. 1985c. U.S. Environmental Protection Agency. U.S. EPA Method 524.1–Volatile organic compounds in water by purge and trap gas chromatography/mass spectrometry. Cincinnati, OH: Environmental Monitoring and Support Laboratory. June 1985 (Revised November 1985.)

U.S. EPA. 1986a. U.S. Environmental Protection Agency. Guidelines for carcinogen risk assessment. *Fed. Reg.* 51(185):33992–34003. September 24.

U.S. EPA. 1986b. U.S. Environmental Protection Agency. Economic evaluation of o-chlorotoluene removal from water by packed column air stripping. Prepared by Office of Water for Health Advisory Treatment Summaries.

Valvani, S.C., S.H. Yalkowsky and T.J. Roseman. 1981. Solubility and partitioning. IV. Aqueous solubility and octanol-water partition coefficients of liquid nonelectrolytes. J. Pharm. Sci. 70(5):502–507.

Van Tassel, S., N. Amalfitano and R.S. Narang. 1981. Determination of arenes and volatile haloorganic compounds in air at microgram per cubic meter levels by gas chromatography. Anal. Chem. 53:2130–2135.

Verschueren, K. 1983. Handbook of Environmental Data on Organic Chemicals, 2nd ed. New York, NY: Van Nostrand Reinhold Co. pp. 386–387.

Wold, J.S. and J.L. Emmerson. 1974. The metabolism of ^{14}C-o-chlorotoluene in the rat. Pharmacologist. 16(2):196. Abstract.

Hexachlorobutadiene

I. General Information and Properties

A. CAS No. 87–68–3

B. Structural Formula

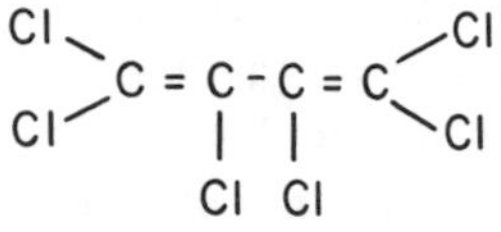

$$\begin{array}{c} Cl \\ \diagdown \\ \diagup \\ Cl \end{array} C = C - C = C \begin{array}{c} Cl \\ \diagup \\ \diagdown \\ Cl \end{array}$$

$$\begin{array}{cc} | & | \\ Cl & Cl \end{array}$$

Hexachlorobutadiene

C. Synonyms

- HCBD, perchlorobutadiene, 1,3-hexachlorobutadiene, 1,1,2,3,4,4-hexachloro-1,3-butadiene (Chemline, 1988).

D. Uses

- Hexachlorobutadiene is used as a solvent in chlorine gas production, an intermediate in the manufacture of rubber compounds and lubricants, a gyroscope fluid and a pesticide (ACGIH, 1980; U.S. EPA, 1980).

E. Properties (Callahan et al., 1979; Banerjee et al., 1980; U.S. EPA, 1980; Hawley, 1981; Ruth, 1986)

Chemical Formula	C_4Cl_6
Molecular Weight	260.7
Physical State (at 25°C)	Clear, colorless liquid
Boiling Point (at 25 mm Hg)	210 to 220°C
Melting Point	–19 to –22°C
Density	—
Vapor Pressure (20°C)	0.15 mm Hg
Specific Gravity (15.5°C)	1.675
Water Solubility (20°C)	Insoluble

Log Octanol/Water
   Partition   3.74
   Coefficient   4.78
Taste Threshold   —
Odor Threshold (air)   12.00 mg/m^3
Conversion Factor   1 ppm = 10.66 mg/m^3
                     1 mg/m^3 = 0.0938 ppm

F. Occurrence

- U.S. EPA (1975) reported that hexachlorobutadiene had been found in low concentrations (< 0.01 µg/L) in domestic drinking water supplies.

- Hexachlorobutadiene was detected at concentrations of 1.9 and 4.7 µg/L in water near Geismar, LA. In a study of the Mississippi delta region, hexachlorobutadiene was found in the water at concentrations < 2 µg/L, but hexachlorobutadiene concentrations in mud or soil were > 200 µg/L. Water samples from an industrial effluent near Geismar, LA, contained < 0.1 to 4.5 µg/L, while levels in the mud were as high as 2,370 µg/L (U.S. EPA, 1976).

- Hexachlorobutadiene was detected in municipal sludges at a median of 4.3 µg/g dry weight from 13 publicly owned treatment works in the United States (Naylor and Loehr, 1982).

G. Environmental Fate

- Because of its low vapor pressure, hexachlorobutadiene may not volatilize rapidly from water to the atmosphere (U.S. EPA, 1980).

- Sorption to soil particles is an important process in the fate of hexachlorobutadiene in water. U.S. EPA (1976) found that hexachlorobutadiene concentrations in the Mississippi delta water were < 2 µg/L, but concentrations in mud or soil were > 200 µg/L. Leeuwangh et al. (1975) found that after initially uncontaminated sediment was allowed to equilibrate with hexachlorobutadiene-contaminated water, the concentration in sediment was 100 times that in the water.

- The half-life of hexachlorobutadiene in soils could not be located in the available literature. However, based on the expected volatility (Callahan et al., 1979) and biodegradability in aquatic media (U.S. EPA, 1981), significant volatilization and biodegradation may not occur in soils (U.S. EPA, 1984a). The compound may, however, be sorbed significantly onto soils containing a high content of organic carbon (U.S. EPA, 1984a). In the absence of significant loss processes, the persistence of hexachlorobutadiene in soil may allow some leaching of the compound into ground water, particularly from sandy soils (U.S. EPA, 1984a).

II. Pharmacokinetics

A. Absorption

- Reichert et al. (1985) administered single gavage doses of 1 or 50 mg/kg ^{14}C-hexachlorobutadiene in a tricaprylin suspension to female Wistar rats (number not specified). Radioactivity recovered 72 hours after treatment indicated that at the 50 mg/kg dose (compared with the 1 mg/kg dose), there was a greater percentage of administered hexachlorobutadiene in the feces (69.03 vs. 42.13%) and decreased renal excretion of metabolites (11.01 vs. 30.61% of the administered dose). The higher amount of radioactivity in the high-dose rat feces was entirely due to unchanged hexachlorobutadiene. At both doses, about 6 to 9% of the dose was recovered in both exhaled air and in carcass and tissues. The authors postulated that the low solubility and high lipophilicity of this compound results in low gastrointestinal absorption and early saturation.

- Nash et al. (1984) administered single gavage doses of ^{14}C-hexachlorobutadiene at 200 mg/kg bw in corn oil to young adult male Wistar-derived rats. Some bile ducts were cannulated, and excreta collected for 5 days. From intact rats, 39% of the administered dose of radioactivity was recovered from the feces, but only 5.4% of the dose was collected from the feces of bile duct cannulated rats. In this experiment, the gastrointestinal absorption from rats treated with a single 200 mg/kg gavage dose was about 95%. Extraction of the gut and its contents indicated that radioactivity in the gut ≥ 16 hours after treatment was associated mainly with water-soluble metabolites and that absorption was "virtually complete" at this time. These results appear to conflict with Reichert et al. (1985), but may be explained by differences in protocols.

- Duprat and Gradiski (1978) reported that dermal doses of 0.25 to 1.0 mL/kg (419 to 1,675 mg/kg) pure hexachlorobutadiene applied to female New Zealand rabbits were completely absorbed in 8 hours.

B. Distribution

- Reichert (1983) and Reichert et al. (1985) observed that liver, brain and kidneys of female Wistar rats contained the highest amounts of radioactivity following single oral doses of 1 or 50 mg ^{14}C-hexachlorobutadiene/

kg bw. However, kidneys contained about twice as much radioactivity covalently bound to tissues as the liver at 1 mg ^{14}C-hexachlorobutadiene/kg bw (Reichert, 1983).

- Nash et al. (1984) administered a single gavage dose of 200 mg ^{14}C-hexachlorobutadiene/kg bw in corn oil to male Wistar-derived rats and sacrificed one rat at 2, 4, 8 and 16 hours. The whole body longitudinal sagittal section autoradiograph from a rat sacrificed 4 hours posttreatment revealed high radioactivity associated with the small intestine and low radioactivity associated with the stomach.

- Jacobs et al. (1974) gave rats oral doses of a mixture of chlorinated hydrocarbons with hexachlorobutadiene at dosages of 2 or 4 mg/kg bw/day for up to 12 weeks. Rats sacrificed at 4, 8 and 12 weeks accumulated about 7 mg hexachlorobutadiene/kg or less in adipose tissue from the inner genital and kidney regions. Lesser quantities of hexachlorobutadiene were found in the liver, heart, kidney and blood. Hexachlorobutadiene did not accumulate in fat to the extent that other chlorinated hydrocarbons in the mixture did. The accumulation of hexachlorobutadiene in body tissues was similar at both dosages.

C. Metabolism

- In rats given single ^{14}C-hexachlorobutadiene gavage doses of 200 mg/kg bw (Nash et al., 1984), direct conjugation with glutathione appeared to be the primary metabolic step. *In vitro* tests with rat liver and kidney preparations performed at the same laboratory confirmed that cytochrome P-450 catalyzed oxidative reactions did not precede conjugation with glutathione (Wolf et al., 1984). The glutathione conjugate formed in rats treated with 200 mg ^{14}C-hexachlorobutadiene/kg was identified as S-(1,1,2,3,4-pentachloro-1:3-butadienyl) glutathione. This metabolite constituted 40% of the radioactivity excreted in the bile collected for 24 hours posttreatment. A metabolite tentatively identified as the cysteinyl glycine conjugate of hexachlorobutadiene, a derivative of the glutathione conjugate, constituted 12% of the radioactivity in the bile. Other radiolabeled metabolites in the bile were not identified.

- Nash et al. (1984) observed substantial enterohepatic recirculation of biliary metabolites in rats given a single oral dose of 200 mg ^{14}C-hexachlorobutadiene/kg bw. The authors propose that hydrolysis and enzymatic degradation of the glutathione conjugate of hexachlorobutadiene and other biliary metabolites lead to the formation of lipophilic moieties that are easily reabsorbed.

- Reabsorbed partially degraded biliary metabolites are further metabolized by renal enzymes, such as β-lyase, to yield a reactive alkylating moiety, probably a thiol, that causes localized kidney damage (Nash et al., 1984; Reichert et al., 1985; Wolf et al., 1984). Other authors have also suggested that this type of metabolic activation is responsible for the observed hexachlorobutadiene nephrotoxicity (Hook et al., 1982; Lock et al., 1984). This proposal is also supported by results of experiments in which oral treatment with hexachlorobutadiene-conjugate produced a pattern of histological and biochemical changes that was "remarkably similar" to that produced by hexachlorobutadiene itself (Lock et al., 1984; Nash et al., 1984). As further support, rats fitted with a biliary cannula were completely protected from kidney damage when dosed orally with hexachlorobutadiene (Nash et al., 1984).

D. Excretion

- After an oral dose of 200 mg ^{14}C-hexachlorobutadiene/kg bw, the principal route of elimination in rats was through bile, with 17 to 20% of the initial dose excreted on each of the first 2 days (Nash et al., 1984). The investigators determined that extensive enterohepatic circulation must have occurred because fecal elimination only amounted to 5% of the total dose of radioactivity/day. Urinary excretion constituted $\leq 3.5\%$ of the total dose/24-hour period.

- Reichert (1983) and Reichert et al. (1985) investigated the elimination of ^{14}C-hexachlorobutadiene following oral administration of 1 or 50 mg/kg bw to rats. After 72 hours, fecal elimination accounted for 42% of the total dose of radioactivity at the low dose and 67% at the high dose. The increase at the higher dose was due entirely to unchanged hexachlorobutadiene. Urinary excretion was 31% of the total at 1 mg/kg and 11% at 50 mg/kg. These results suggested to the investigators that gastrointestinal absorption is saturable at higher doses. Elimination in expired air was consistent, with about 5% eliminated as unchanged hexachlorobutadiene at both doses and 1 or 3% as CO_2 at the high and low dose, respectively.

III. Health Effects

A. Humans

1. Short-term Exposure

- Data regarding short-term exposure of humans to hexachlorobutadiene were not located in the available literature.

2. Long-term Exposure

- According to an abstract, Krasniuk et al. (1969) examined 153 farm workers exposed intermittently for 4 years to soil and grape fumigants containing hexachlorobutadiene and polychlorobutane-80. Compared with 52 unexposed workers, the exposed workers exhibited a higher incidence of arterial hypotension, myocardial dystrophy, nervous disorders (sleep disorders, nausea, hand trembling), disordered smell functions, liver effects, chest pains and upper respiratory tract changes.

B. Animals

1. Short-term Exposure

- Schwetz et al. (1977) reported single-dose oral LD_{50} values of 46 and 65 mg/kg for 21-day-old female and male rats, respectively. Among adult rats, values of 200 to 400 mg/kg were reported for females and 504 to 667 mg/kg for males. U.S. EPA (1980) reported single dose oral LD_{50} values of 90 mg/kg for guinea pigs and 80 to 116 mg/kg for mice.

- In a series of experiments, Alderly Park (Wistar-derived) albino rats and Fisher 344 rats were administered single interperitoneal (i.p.) doses of hexachlorobutadiene ranging from 25 to 400 mg/kg. While renal efflux rates of the organic anion, p-aminohippurate (PAH) and the organic cation, tetraethylammonium (TEA) were unaffected, accumulation of PAH was decreased while that of TEA was unaffected suggesting damage to the renal organic cation uptake system (Hook et al., 1982; Kuo and Hook, 1983). Kidney-to-body weight ratios, but not liver-to-body weight ratios, were increased in all groups; and increases in blood urea nitrogen were noted for all groups, but the effect was more pronounced in younger rats (Kuo and Hook, 1983). Adult male rats were less susceptible to hexachlorobutadiene-induced nephrotoxicity than adult female and young male rats; the authors proposed that the age and sex differences were due to differences in hepatic and renal enzymes responsible for activation or detoxication of hexachlorobutadiene (Hook et al., 1983).

- Harleman and Seinen (1979) administered diets containing 0, 50, 150 or 450 ppm hexachlorobutadiene to groups of six male and six female weanling Wistar rats for 14 days. Histopathological changes in kidneys occurred in all hexachlorobutadiene-treated animals. These consisted of dose-related degeneration of tubular epithelial cells. No histological effects were observed in the liver, the only other organ that was examined. Relative kidney weights were significantly increased in rats at the two highest doses. Body weight gain was also significantly depressed in treated rats in a dose-related manner. Based on body weight and food consumption data provided, actual hexachlorobutadiene consumption in the low-dose (50 ppm) group was 4.66 and 4.447 mg/kg/day for males and females, respectively.

- Kociba et al. (1971) reported the results of an unpublished Dow Chemical Co. study in which female weanling Sprague-Dawley rats (4/group) received hexachlorobutadiene in the diet at dosages of 0, 1, 3, 10, 30, 65 or 100 mg/kg bw/day for 30 days. Increased relative kidney weight and renal tubular necrosis occurred at $\geq$ 30 mg/kg/day. Increased hemoglobin concentration, decreased food consumption and reduced body weight gain occurred at dosages $\geq$ 10 mg/kg/day. A "marginal but statistically significant" increase in relative kidney weight was reported at 3 mg/kg/day but not at 10 mg/kg/day.

2. Dermal/Ocular Effects

- Duprat and Gradiski (1978) reported a rabbit dermal LD_{50} of 0.72 mL/kg in 8-hour exposures. Cutaneous necrosis occurred at the site of application. Signs of systemic toxicity were renal epithelial necrosis and fatty liver degeneration.

- Gradiski et al. (1975) found that a 10% hexachlorobutadiene solution (solvent not stated) caused only slight dermal and ocular mucosa irritation in rabbits. Guinea pigs exhibited delayed allergic reactions to dermal hexachlorobutadiene application.

3. Long-term Exposure

- Harleman and Seinen (1979) conducted a study in which groups of 10 male and 10 female weanling Wistar rats were given gavage doses of 0, 0.4, 1.0, 2.5, 6.3 or 15.6 mg/kg bw/day for 13 weeks. Degeneration of proximal renal tubules occurred at doses of 2.5 and $\geq$ 6.3 mg/kg in females and males, respectively. Urine concentrating ability was significantly reduced in females at $\geq$ 2.5 mg/kg/day and in males at 15.6 mg/kg. Liver weights were increased in females at 15.6 mg/kg bw and in males at 6.3 and 15.6 mg/kg. Increased cytoplasmic basophilia of hepatocytes associated with the increased liver weight occurred in males at the two highest doses. Increased relative kidney weights and growth inhibition occurred at the two highest doses in both males and females. A slight increase in relative kidney weights occurred in males at the lower doses which was not considered significant by the authors. The authors concluded that 1.0 and 2.5 mg/kg bw/day were NOAELs for females and males, respectively, in this study.

- In a 148-day reproduction study (Schwetz et al., 1977), groups of 10 to 17 male and 20 to 34 female adult rats were fed diets containing hexachlorobutadiene that provided dosages of 0, 0.2, 2.0 or 20 mg/kg bw/day for 90 days before mating, 15 days during mating, and subsequently throughout gestation (22 days) and lactation (21 days). Signs of toxicity occurred at the two higher dosages and included weight gain and histological changes in kidneys and decreased food consumption (only at 20 mg/kg bw/day in males). At the highest dose, increases were observed in relative kidney weight (both sexes) and relative liver weight (males only). No observable toxic effects occurred at 0.2 mg/kg bw/day.

- Kociba et al. (1977) conducted a chronic study in which groups of Sprague-Dawley rats received diets containing hexachlorobutadiene at dosages of 0. 0.2. 2.0 or 20 mg/kg bw/day for 2 years. Group sizes were 39 to 40 rats/sex/dose level for treated groups and 90 rats/sex for controls. Urinalyses, clinical chemistry analyses, hematological analyses and histological examinations of major organs were performed. Ingestion of 20 mg/kg bw/day resulted in increased urinary excretion of coproporphyrin, increased kidney weight and increased renal tubular epithelial hyperplasia. High-dose rats also experienced decreased body weight gain and decreased survival (males only). At 2 mg/kg bw/day, there was an increase in urinary coproporphyrin excretion (females only) and an increase in renal tubular epithelial hyperplasia/proliferation. No observable treatment-related adverse effects occurred at 0.2 mg/kg bw/day. As in short-term studies, the kidney appeared to be the primary target organ (Kociba et al., 1977).

4. Reproductive Effects

- Schwetz et al. (1977) conducted a 148-day study in which groups of 10 to 17 male and 20 to 34 female adult rats/group were fed diets containing hexachlorobutadiene at dosages of 0, 0.2, 2.0 or 20 mg/kg bw/day for 90 days prior to mating, 15 days during mating, and subsequently throughout gestation (22 days) and lactation (21 days). There were no treatment-related effects on pregnancy or neonatal survival. Body weight of 21-day-old weanlings in the high-dose group was slightly but significantly ($p < 0.05$) less than controls. No toxic effects were observed in neonates at doses of 0.2 or 2.0 mg/kg/day. Signs of toxicity occurred in adults at the two higher dosages and included histological kidney changes and decreased food consumption and weight gain.

- Harleman and Seinen (1979) administered 0, 150 or 1500 ppm of hexachlorobutadiene in the diet to groups of 6 female, 10-week-old Wistar rats. These doses are equivalent to 0, 15 or 150 mg/kg/day, respectively, assuming a body weight of 0.1 kg and a food consumption rate of 0.01 kg/day (Lehman, 1959). The rats were mated after the fourth week of administration. No conception was observed in the 1500 ppm dosage group. There was no significant difference in the number of offspring or in the resorption rate at the 150 ppm dosage.

5. Developmental Effects

- No treatment-related effects on neonatal development were observed in the Schwetz et al. (1977) study described under Reproductive Effects.

- Badaeva (1983) reported that oral doses of 8.1 mg hexachlorobutadiene/ kg bw/day given to pregnant female rats throughout gestation resulted in offspring having ultrastructural changes in neurocytes and higher levels of free radicals in the brain and spinal cord. Offspring of treated rats also had lower body weights and shorter crown-rump lengths than controls.

- Harris et al. (1979) reported that pups of rats injected intraperitoneally with 10 mg hexachlorobutadiene/kg bw/day on days 1 to 15 of gestation experienced three times as many soft tissue anomalies as controls, although no particular type of anomaly was predominant.

- Harleman and Seinen (1979) administered 0, 150 or 1500 ppm of hexa-chlorobutadiene in the diet to groups of 6 female, 10-week-old Wistar rats. These doses are equivalent to 0, 15 or 150 mg/kg/day, respectively, assuming a body weight of 0.1 kg and a food consumption rate of 0.01 kg/day (Lehman, 1959). The rats were mated after the fourth week of administration. Pups had lower birth weights and reduced growth compared to controls. No gross malformations were observed.

6. Mutagenicity

- Data regarding the mutagenicity of hexachlorobutadiene are conflicting.

- Several authors have reported that hexachlorobutadiene was not mutagenic in the Ames assay with *Salmonella typhimurium*, either with or without a mammalian metabolic activation system (Rapson et al., 1980; Reichert et al., 1983; Stott et al., 1981; DeMeester et al., 1981).

- Reichert et al. (1984) reported that hexachlorobutadiene was mutagenic in *S. typhimurium* with an S-9 activation system, and Simmon (1977) reported hexachlorobutadiene to be a potent mutagen in *S. typhimurium* TA100.

- Metabolites and derivatives of hexachlorobutadiene were mutagenic in *S.typhimurium* with metabolic activation (Wild et al., 1986; Green et al., 1983; Reichert and Schutz, 1986) and without metabolic activation (Dekant et al., 1986).

- Schiffmann et al. (1984) reported that hexachlorobutadiene caused an increase in unscheduled DNA synthesis in Syrian hamster embryo fibroblasts both in the presence and absence of an exogenous metabolizing system. Morphological transformation was also induced.

- Woodruff et al. (1985) reported that hexachlorobutadiene was not mutagenic in *Drosophila* either by feeding or by adult injection.

- Treatment of rats with oral doses of hexachlorobutadiene (20 mg/kg bw/day) for 3 weeks resulted in cytotoxic effects (a 1.8 times increase in renal DNA synthesis) and genotoxic effects (renal DNA alkylation and a 1.4 times increase in renal DNA repair) (Stott et al., 1981).

7. Carcinogenicity

- Kociba et al. (1977) found that administration of hexachlorobutadiene in the diet at dosages of 20 mg/kg/day for 2 years caused renal tubular adenomas and adenocarcinomas in rats. Incidences of renal tubular neoplasms in males and females were 23.1% and 15%, respectively, compared with 1.1% and 0% in controls. In two cases, the renal tumors metastasized in the lungs. No renal tubular neoplasms were observed in rats ingesting 2.0 or 0.2 mg/kg/day. No other treatment-related neoplasms occurred. The authors concluded that hexachlorobutadiene-induced renal neoplasms occurred only at doses higher than those causing discernible renal injury.

- Theiss et al. (1977) investigated the induction of lung adenomas in male strain A mice following the i.p. administration of hexachlorobutadiene (4 or 8 mg/kg bw) 3 times/week until totals of 52 or 96 mg were injected. There was no statistically significant increase in the mean number of lung tumors per test mouse as compared with vehicle-treated controls, nor was a dose-response relationship obtained.

- Chudin et al. (1985) reported that oral hexachlorobutadiene doses of 0.6, 5.8 or 37 mg/kg bw/day for 1 year were not carcinogenic in rats, although some benign tumors of the kidney and liver were noted.

- Van Duuren et al. (1979) reported that hexachlorobutadiene did not act as an initiator in an initiation/promotion experiment in mouse skin, nor did it cause tumors in the skin or distant organs after repeated application to the skin.

IV. Quantification of Toxicological Effects

A. One-day Health Advisory

No oral studies are suitable for calculation of a One-day HA. For this reason, the Ten-day HA of 0.3 mg/L is recommended as a conservative estimate for use as the One-day HA.

B. Ten-day Health Advisory

The study by Kociba et al. (1971) has been selected to serve as the basis for the 10-kg child Ten-day HA because it defines a NOAEL for histopathological changes in the kidney, the most sensitive target organ, for rats in a 30-day exposure. Adverse kidney effects including renal tubular degeneration and necrosis and increased relative kidney weight occurred at dosages of $\geq$ 30 mg hexachlorobutadiene/kg bw/day. A "marginal but statistically significant" increase in relative kidney weight occurred at 3 mg/kg bw/day, presumably without histopathological effects. It therefore seems appropriate to consider 3 mg/kg/day a NOAEL for kidney histopathology in rats in this study. This NOAEL is lower than dosages (4.5 mg/kg bw/day) that caused histopathological kidney changes in the 14-day rat dietary study by Harleman and Seinen (1979).

The Ten-day HA for the 10-kg child is calculated as follows:

$$\text{Ten-day HA} = \frac{(3 \text{ mg/kg/day}) (10 \text{ kg})}{(100) (1 \text{ L/day})} = 0.3 \text{ mg/L} (300 \text{ } \mu g/L)$$

C. Longer-term Health Advisory

The 13-week gavage study by Harleman and Seinen (1979) has been selected to serve as the basis for the Longer-term HA because it defines a NOAEL for histopathological kidney effects in rats that is the highest NOAEL below which no adverse effects have been reported in studies of appropriate duration. In this study, histopathological and functional effects on kidneys occurred at dosages of $\geq$ 2.5 mg/kg bw/day. A slight increase in relative kidney weights occurred in males at the lower doses which was not considered significant by the authors. Therefore, a dosage of 1 mg/kg bw/day is considered a NOAEL. Supporting data were provided by the 148-day dietary rat study by Schwetz et al. (1977), in which histopathological effects in kidneys occurred at 2.0 but not 0.2 mg/kg bw/day.

The Longer-term HA for a 10-kg child is calculated as follows:

$$\text{Longer-term HA} = \frac{(1 \text{ mg/kg/day}) (10 \text{ kg})}{(100) (1 \text{ L/day})} = 0.1 \text{ mg/L} (100 \text{ } \mu g/L)$$

The Longer-term HA for a 70-kg adult is calculated as follows:

$$\text{Longer-term HA} = \frac{(1 \text{ mg/kg/day}) \ (70 \text{ kg})}{(100) \ (2 \text{ L/day})} = 0.35 \text{ mg/L [rounded to}$$
$$4 \text{ mg/L} \ (400 \ \mu\text{g/L})]$$

D. Lifetime Health Advisory

The study by Kociba et al. (1977) is the most appropriate from which to derive the DWEL. In this study, rats were fed diets that provided hexachlorobutadiene at 0, 0.2, 2.0 or 20 mg/kg/day for 2 years. Functional and histopathological effects occurred in the kidneys of rats at $\geq$ 2 mg/kg/day. No effects were reported at 0.2 mg/kg/day.

From these results, a NOAEL of 0.2 mg/kg/day was identified. Using this NOAEL, the DWEL is derived as follows:

Step 1: Determination of the Reference Dose (RfD)

$$\text{RfD} = \frac{(0.2 \text{ mg/kg/day})}{(100)} = 0.002 \text{ mg/kg/day}$$

Step 2: Determination of the Drinking Water Equivalent Level (DWEL)

$$\text{DWEL} = \frac{(0.002 \text{ mg/kg/day}) \ (70 \text{ kg})}{(2 \text{ L/day})} = 0.07 \text{ mg/L} \ (700 \ \mu\text{g/L})$$

Step 3: Determination of the Lifetime Health Advisory

$$\text{Lifetime HA} = \frac{(0.07 \text{ mg/kg/day}) \ (20\%)}{(10)} = 0.0014 \text{ mg/L [rounded to}$$
$$0.001 \text{ mg/L} \ (1 \ \mu\text{g/L})]$$

E. Evaluation of Carcinogenic Potential

- IARC has not classified hexachlorobutadiene in terms of its carcinogenic potential, although IARC (1979) stated that there was "limited evidence" that hexachlorobutadiene was carcinogenic in rats.

- The U.S. EPA derived an upper limit for the carcinogenic potency factor (q_1*) of 7.8×10^{-2} using the incidences of renal tubular adenomas and carcinomas in rats (Kociba et al., 1977). Using this q_1* and assuming an average water consumption of 2 L/day and a human reference body weight of 70 kg, the water concentrations associated with the 95% upper limit lifetime 10^{-4}, 10^{-5} and 10^{-6} risk levels are calculated to be 50, 5 and 0.5 μg/L, respectively. Applying the criteria described in the U.S. EPA's Guidelines for Carcinogen Risk Assessment (U.S. EPA, 1986b), hexachlorobutadiene has been classified in Group C: possible human carcinogen. This category is for agents with limited evidence of carcinogenicity in animals in the absence of human data.

V. Other Criteria, Guidance and Standards

- ACGIH (1980, 1985) recommended a TLV of 0.02 ppm (0.24 mg/m^3) for hexachlorobutadiene and classified it as a suspected carcinogen. U.S. EPA (1980) calculated levels resulting in incremental increase in lifetime cancer risk of 10^{-5}, 10^{-6} and 10^{-7}. The corresponding recommended criteria were 4.47 μg/L, 0.45 mg/L and 0.045 μg/L, respectively. If these estimates were made for the consumption of aquatic organisms only and excluding consumption of water, the levels would be 500, 50 and 5.0 μg/L, respectively.

VI. Analytical Methods

- Analysis of hexachlorobutadiene is by a gas chromatographic (GC) method applicable to the determination of certain chlorinated hydrocarbons in water samples (U.S. EPA, 1984b). In this method, about 1 L of sample is extracted with methylene chloride using a separatory funnel. The methylene chloride extract is dried and exchanged to hexane during concentration to a volume of $\leq$ 10 mL. The extract is separated by GC and the compounds are measured with an electron capture detector. The method detection limit has been estimated at 0.34 μg/L for hexachlorobutadiene. Confirmatory analysis may be done with a second GC column or by mass spectrometry (U.S. EPA. 1984c).

VII. Treatment Technologies

- Thiem et al. (1987) report the results from a pilot designed to simulate conventional water treatment followed by PAC (powdered activated carbon) and GAC (granular activated carbon) adsorption for the removal of shock loadings of specified organic chemicals, including hexachlorobutadiene. The contaminant influent concentration was varied from 2, 10, 100 to 200 μg/L. PAC removal effectiveness was studied at three different dosages: 2, 20 and 50 mg/L. In general, there was a higher removal efficiency, with a higher hexachlorobutadiene influent concentration. PAC capacity for hexachlorobutadiene was calculated in the range of 48 to 60 mg/mg of carbon at a PAC dosage of 2 mg/L and a contaminant concentration of 200 μg/L, while PAC adsorption capacity of 9 to 10 mg/mg carbon was calculated at a PAC dose of 20 mg/L at the 200 μg/L influent concentration.

- Van Dyke et al. (1986) studied and reported the efficiency of a home-use water filter containing pressed carbon block as a filtering media for the removal of a number of organic chemicals, including hexachlorobutadiene. The filtering system consisted of a nonwoven prefilter, a pressed carbon block, and a porous polyethylene-fritted core. The water was supplied at a constant pressure of 50 psig. Each run consisted of passing a volume of water equal to 150 percent of the filter rated life of 500 gallons, and analyzing for the various contaminants. Hexachlorobutadiene was present in the influent at a concentration of 20 μg/L. This system removed hexachlorobutadiene below its detection limit (0.1 μg/L).

- U.S. EPA (1986a) estimated the feasibility of removing hexachlorobutadiene from water by packed column aeration employing the engineering design procedure and cost model presented at the 1983 national ASCE Conference on Environmental Engineering. Based on chemical and physical properties and assumed operating conditions, a 90% removal efficiency of hexachlorobutadiene was postulated for a column with a diameter of 5.5 feet and packed with 12 feet of 1-inch plastic saddles. The air-to-water ratio required to achieve this degree of removal effectiveness is 6:1.

VIII. References

ACGIH. 1980. American Conference of Governmental Industrial Hygienists. Documentation of the threshold limit values. 4th ed. Cincinnati, OH: ACGIH. pp. 2ll-212.

ACGIH. 1985. American Conference of Governmental Industrial Hygienists. Threshold limit values for chemical substances in the work environment. Adopted by ACGIH with intended changes for 1985–1986. Cincinnati, OH: ACGIH. p. 20.

Badaeva, L.N. 1983. Structural and metabolic indexes of the postnatal neurotoxicity of organochloride pesticides. Dopov. Akad. Nauk Ukr. pp. 55–58. (CA 099:083463W).

Banerjee, S., S.H. Yalkowsky and S.C. Valvani. 1980. Water solubility and octanol/water partition coefficients of organics. Limitations of the solubility-partition coefficient correlation. Environ. Sci. Technol. 14:1227–1229.

Callahan, M.A., M.W. Slimak, N.W. Gabel et al. 1979. Water-related environmental fate of 129 priority pollutants, Vol. II. EPA-440/4–79–029B. Washington, DC: U.S. EPA. (Cited in U.S. EPA, 1984.)

Chemline. 1988. Online. Bethesda, MD: National Library of Medicine.

Chudin, V.A., Z.A. Gafieva, N.A. Koshurnikova, A.P. Nifatov and V.S. Revina. 1985. Evaluating the mutagenicity and carcinogenicity of hexachlorobutadiene. Gig. Sanit. pp. 79–80. (CA 103:033344P).

Dekant, W., S. Vamvakas, K. Berthold, S. Schmidt, D. Wild and D. Henschler. 1986. Bacterial β-lyase mediated cleavage and mutagenicity of cysteine conjugates derived from the nephrocarcinogenic alkenes trichloroethylene, tetrachloroethylene and hexachlorobutadiene. Chem.-Biol. Interactions. 60:31–45.

DeMeester et al. 1981. Mutagenic activity of butadiene, hexachlorobutadiene and isoprene. In: Ind. Environ. Xenobiotics Proc. Int. Conf. pp. 195–203.

Duprat, P. and D. Gradiski. 1978. Percutaneous toxicity of hexachlorobutadiene. Acta Pharmacol. Toxicol. 43(5):346–353.

Fitch, J.H., Jr. 1985. Ion exchange potentially effective in reducing effluent from bleach plant. Pulp Pap. 59(4):132–135.

Gradiski, D., P. Duprat, J.L. Magadur and E. Fayein. 1975. Toxicological and experimental study of hexachlorobutadiene. J. Eur. Toxicol. 8(3):180–187. (PESTAB 76: 1531).

Green, T., J. Nash, J. Odum and E.F. Howard. 1983. The renal metabolism of a glutathione conjugate of the carcinogen hexachloro-1:3-butadiene: evidence for the formation of a mutagenic metabolite in the rat kidney. In: Extrahepatic Drug Metabolism and Chemical Carcinogenesis. New York, NY: Elsevier Science Publishers. pp.623–624.

Harleman, J.R. and W. Seinen. 1979. Short-term toxicity and reproduction studies in rats with hexachloro-(1,3)-butadiene. Toxicol. Appl. Pharmacol. 47:1–14.

Harris, S.J., G.P. Bond and R.W. Niemeier. 1979. The effects of 2-nitropropane, naphthalene and hexachlorobutadiene on fetal rat development. Toxicol. Appl. Pharmacol. 43:A35. Abstract.

Hawley, G.G. 1981. The Condensed Chemical Dictionary. 10th ed. New York, NY: Van Nostrand Reinhold Co. p. 26.

Hook, J.B., J. Ishmael and E.A. Lock. 1983. Nephrotoxicity of hexachloro-1:3-butadiene in the rat: the effect of age, sex and strain. Toxicol. Appl. Pharmacol. 67(1):122–131.

Hook, J.B., M.S. Rose and E.A. Lock. 1982. The nephrotoxicity of hexachloro-1:3-butadiene in the rat: studies of organic anion and cation transport in renal slices and the effect of monooxygenase inducers. Toxicol. Appl. Pharmacol. 65(3):373–382.

IARC. 1979. International Agency for Research on Cancer. IARC monographs on the evaluation of the carcinogenic risk of chemicals to humans. Hexachlorobutadiene. Lyon. France: IARC, WHO. 20:179–189.

Jacobs, A. et al. 1974. Accumulation of noxious chlorinated substances from

Rhine River water in the fatty tissue of rats. Vom. Wasser. 43:259. (Cited in U.S. EPA, 1980.)

Kociba, R.J., P.J. Gehring, C.G. Humiston and G.L. Sparschu. 1971. Toxicologic study of female rats administered hexachlorobutadiene or hexachlorobenzene for thirty days. Midland, MI: Dow Chemical Co. (Cited in Schwetz et al., 1977; U.S. EPA, 1980.)

Kociba, R.J., D.G. Keyes, G.C. Jersey et al. 1977. Results of a two-year chronic toxicity study with hexachlorobutadiene in rats. J. Am. Ind. Hyg. Assoc. 38:589–602.

Krasniuk, E.P., L.A. Zaritskaia, V.G. Boiko, G.A. Voitenko and L.A. Matokhniuk. 1969. The state of health of viticulturists having contact with the fumigants, hexachlorobutadiene and polychlorobutane-80. Vrachebnoe Delo No. pp. 111–115. (HAPAB 70:01900).

Kuo, C.R. and J.B. Hook. 1983. Effects of age and sex on hexachloro-1,3-butadiene toxicity in the Fischer 344 rat. Life Sci. 33(6):517–523.

Leeuwangh, P. et al. 1975. Toxicity of hexachlorobutadiene in aquatic organisms: sublethal effects of toxic chemicals on aquatic animals. In: Proc. of Swedish-Netherlands Symp., September 2–5. New York, NY: Elsevier Scientific Publishing Co., Inc. (Cited in U.S. EPA, 1980.)

Lehman, A.J. 1959. Appraisal of the safety of chemicals in foods, drugs, and cosmetics. Q. Bull. Assoc. Food Drug Office. Topeka, KS: U.S. Association of Food and Drug Officials of the United States.

Lock, E.A., J. Ishmael and J.B. Hook. 1984. Nephrotoxicity of hexachloro-1,3-butadiene in the mouse: the effect of age, sex, strain, monooxygenase modifiers, and the role of glutathione. Toxicol. Appl. Pharmacol. 72(3):484–494.

Nash, J.A., L.J. King, E.A. Lock and T. Green. 1984. The metabolism and disposition of hexachloro-1:3-butadiene in the rat and its relevance to nephrotoxicity. Toxicol. Appl. Pharmacol. 73:124–137.

Naylor, L.M. and R.C. Loehr. 1982. Priority pollutants in municipal sewage sludge. Biocycle. 23:18–22.

Perry, R.H. and C.H. Chilton. 1973. Chemical Engineers Handbook, 5th ed. New York, NY: McGraw Hill Book Company.

Rapson, W.B., M.A. Nazar and V.V. Butsky. 1980. Mutagenicity produced by aqueous chlorination of organic compounds. Bull. Environ. Contam. Toxicol. 24:590–596.

Reichert, D. 1983. Metabolism and disposition of hexachloro(1,3)butadiene in rats. Dev. Toxicol. Environ. Sci. 11:411–414.

Reichert, D., T. Neudecker, U. Spengler and D. Henschler. 1983. Mutagenicity of dichloroacetylene and its degradation products trichloroacetyl chloride, trichloroacryloyl chloride and hexachlorobutadiene. Mutat. Res. 117:21–29.

Reichert, D., T. Neudecker and S. Schutz. 198_
butadiene, perchlorobutenoic acid and pe_
Mutat. Res. 137(2–3):89–93.

Reichert, D., S. Schutz and M. Metzler. 1985. E_
lism of hexachlorobutadiene in rats. Evidence t_
conjugation reactions. Biochem. Pharmacol. 34(4_

Reichert, D. and S. Schutz. 1986. Mercapturic acid forr_
and intermediary step in the metabolism of hexachlorou_
Pharmacol. 35(8):1271–1275.

Ruth, J.H. 1986. Odor thresholds and irritation levels of s_
substances: a review. J. Am. Ind. Hyg. Assoc. 47:A-142 to /_

Schiffman, D., D. Reichert and D. Henschler. 1984. Induction o_
logical transformation and unscheduled DNA synthesis in Syrian
embryo fibroblasts by hexachlorobutadiene and its putative metaboli_
tachlorobutenoic acid. Cancer Lett. 23(3):297–305.

Schwetz, B.A., F.A. Smith, C.G. Humiston, J.F. Quast and R.J. Kocib_
1977. Results of a reproduction study in rats fed diets containing hexachlorobutadiene. Toxicol. Appl. Pharmacol. 42:387–398.

Simmon, V.F. 1977. Structural correlation of carcinogenic and mutagenic
alkyl halides. In: Proc., 2nd FDA Office of Science Summer Symposium,
U.S. Naval Academy, August 31 to September 2. pp. 163, 171. (Cited in
Stott et al., 1981.)

Sittig, M. 1985. Handbook of Toxic and Hazardous Chemicals and Carcinogens, 2nd ed. Park Ridge, NJ: Noyes Publications.

Stott, W.T., J.F. Quast and P.G. Watanabe. 1981. Differentiation of the
mechanisms of oncogenicity of 1,4-dioxane and 1,3-hexachlorobutadiene in
the rat. Toxicol. Appl. Pharmacol. 60(2):287–300.

Theiss, J.C., G.D. Stoner, M.B. Shimkin and E.K. Weisburger. 1977. Tests
for carcinogenicity of organic contaminants of United States drinking
waters by pulmonary tumor response in strain A mice. Cancer Res.
37:2717–2720.

Thiem, I.T., D.L. Badorek, A. Johari and E. Alkhatib. 1987. Absorption of
synthetic organic shock loadings. J. Environ. Eng. 113(6):1302–1318.

U.S. EPA. 1975. U.S. Environmental Protection Agency. Preliminary assessment of suspected carcinogens in drinking water. EPA 560/4-75-003.
Washington, DC: Office of Toxic Substances. (Cited in U.S. EPA, 1980.)

U.S. EPA. 1976. U.S. Environmental Protection Agency. An ecological study
of hexachlorobutadiene. EPA 560/6-76-010. Washington, DC: Office of
Toxic Substances. (Cited in U.S. EPA, 1980.)

U.S. EPA. 1980. U.S. Environmental Protection Agency. Ambient water quality criteria for hexachlorobutadiene. Prepared by the Office of Health and
Environmental Assessment, Environmental Criteria and Assessment

, Cincinnati, OH for the Office of Water Regulations and Standards, ington, DC. EPA 440/5-80-53. NTIS PB 81-117640.

PA. 1981. U.S. Environmental Protection Agency. Aquatic fate pro-ses data for organic priority pollutants. EPA 440/4-81-014. Washing-n, DC: Monitoring and Data Support Division, Office of Water Regula-ons and Standards.

. EPA. 1984a. U.S. Environmental Protection Agency. Health effects assessment for hexachlorobutadiene. Prepared by the Office of Health and Environmental Assessment, Environmental Criteria and Assessment Office, Cincinnati, OH for the Office of Emergency and Remedial Response, Washington, DC. EPA 540/1-86-053.

U.S. EPA. 1984b. U.S. Environmental Protection Agency. U.S. EPA Method 612-Chlorinated hydrocarbons. *Fed. Reg.* 49(209):128-135. October 24.

U.S. EPA. 1984c. U.S. Environmental Protection Agency. U.S. EPA Method 625-Base/neutral and acids. *Fed. Reg.* 49(209):153-174. October 26.

U.S. EPA. 1986a. U.S. Environmental Protection Agency. Economic evaluation of hexachlorobutadiene removal from water by packed column air stripping. Prepared by Office of Water for Health Advisory Treatment Summaries.

U.S. EPA. 1986b. U.S. Environmental Protection Agency. Guidelines for carcinogen risk assessment. *Fed. Reg.* 51(185):33992-34003. September 24.

Van Duuren, B.L., B.M. Goldschmidt, G. Lowengart et al. 1979. Carcinogenicity of halogenated olefinic and aliphatic hydrocarbons in mice. J. Natl. Cancer Inst. 63(6):1433-1439.

Van Dyke, K., R. Kuennen, J. Stiles, J. Wezeman and J. O'Neal. 1986. Test stand design and testing for a pressed carbon block water filter. American Laboratory. 18(9):118-132.

Verschueren, K. 1977. Handbook of Environmental Data on Organic Chemicals. New York, NY: Van Nostrand Reinhold Co.

Wild, D., S. Schultz and D. Reichert. 1986. Mutagenicity of the mercapturic acid and other S-containing derivatives of hexachloro-1,3-butadiene. Carcinogenesis. 7(3):431-434.

Wolf, C.R., P.N. Berry, J.A. Nash, T. Green and E.A. Lock. 1984. Role of microsomal and cytosolic glutathione S-transferases in the conjugation of hexachloro-1:3-butadiene and its possible relevance to toxicity. J. Pharmacol. Exp. Thera. 228:202-208.

Woodruff, R.C., J.M. Mason, R. Valencia and S. Zimmering. 1985. Chemical mutagenesis testing in *Drosophila*. V. Results of 53 coded compounds tested for the National Toxicology Program. Environ. Mutagen. 7(5):677-702.

1,1,1,2-Tetrachloroethane

I. General Information and Properties

A. CAS No. 630–20–6

B. Structural Formula

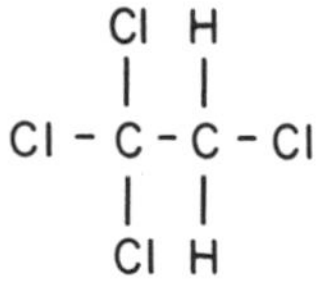

$$\begin{array}{c} \quad\;\; Cl \;\; H \\ \quad\;\; | \quad | \\ Cl - C - C - Cl \\ \quad\;\; | \quad | \\ \quad\;\; Cl \;\; H \end{array}$$

1,1,1,2-Tetrachloroethane

C. Synonyms

- Synonyms could not be located for 1,1,1,2-tetrachloroethane.

D. Uses

- 1,1,1,2-Tetrachloroethane is used as a feedstock for the production of solvents such as trichloroethylene and tetrachloroethylene (Archer, 1979).

E. Properties (Dilling, 1977; Archer, 1979; HSDB, 1988)

Chemical Formula	CCl_3CH_2Cl
Molecular Weight	167.9
Physical State	Liquid
Boiling Point	130.5°C
Melting Point	–68.7°C
Density (20°C)	1.54064 g/mL
Vapor Pressure (25°C)	13.9 mm Hg
Specific Gravity (20°C)	1.54064 g/mL
Water Solubility (25°C)	1,100 mg/L
Log Octanol/Water	
Partition	—
Coefficient	—

69

Taste Threshold	—
Odor Threshold	—
Conversion Factor	1 ppm = 6.87 mg/m³
	1 mg/m³ = 0.1455 ppm

F. Occurrence

- 1,1,1,2-Tetrachloroethane was identified in finished drinking water in the New Orleans area at concentrations of 0.04 to 0.11 µg/L (Keith et al., 1976) and in the Washington, D.C. water supply at a concentration of 1 µg/L (Scheiman et al., 1974).

- Mean atmospheric concentrations of 1,1,1,2-tetrachloroethane in Los Angeles, CA; Phoenix, AZ; and Oakland, CA were 3.7, 8.5 and 4.2 ppt (25, 58 and 29 ng/m³), respectively (Singh et al., 1981).

- Only two locations in a nationwide survey showed detectable atmospheric concentrations of 1,1,1,2-tetrachloroethane (U.S. EPA, 1978). Concentrations in Front Royal, VA and the Houston-Pasadena-Deer Park-Freeport-LaPorte area were 810 and 21 µg/m³, respectively.

G. Environmental Fate

- Evaporation is likely to be the predominant fate determining process for aqueous 1,1,1,2-tetrachloroethane. Dilling (1977) determined a half-life of 42.3 minutes for evaporation from a 0.9 ppm aqueous solution (6.5 cm depth, 200 rpm stirring, 25°C still air).

- Significant chemical or microbial degradation of 1,1,1,2-tetrachloroethane is not expected based on observations or predictions of the fate of 1,1,2,2-tetrachloroethane (U.S. EPA, 1979; Tabak et al., 1981; Pearson and McConnell, 1975).

II. Pharmacokinetics

A. Absorption

- No data specifically addressing the absorption kinetics of 1,1,1,2-tetrachloroethane were found in the available literature. Truhaut et al. (1974), however, found systemic effects following oral (800 to 2,500 mg/kg), inhalation (800 to 4,200 ppm) or dermal (11 to 22 g/kg) administration of 1,1,1,2-tetrachloroethane in rats, mice and rabbits, indicating that absorption does occur by these routes.

B. Distribution

- Data regarding the distribution of 1,1,1,2-tetrachloroethane could not be located in the available literature.

C. Metabolism

- Truhaut (1972) identified 2,2,2-trichloroethanol and trichloroethyl glucuronic acid in the urine of rats, guinea pigs and rabbits after oral administration of 1,1,1,2-tetrachloroethane (vehicle not specified). Trichloroethanol was further oxidized in rats to trichloroacetic acid, which was also found in the urine.

- Ikeda and Ohtsuji (1972) exposed male and female Wistar rats to 1,1,1,2-tetrachloroethane by inhalation (1,373 mg/m^3) or intraperitoneal (i.p.) injection (467 mg/kg). The major urinary metabolites for both exposure routes were trichloroethanol and trichloroacetic acid. The authors proposed a pathway involving hydrolytic dehalogenation to trichloroethanol, and further oxidation to trichloroacetic acid.

- Thompson et al. (1984) found high levels of dichloroethylene in the blood of male Sprague-Dawley rats pretreated with phenobarbital and subsequently intraperitoneally injected with 1,1,1,2-tetrachloroethane.

- Yllner (1971) found that subcutaneous administration of 1,1,1,2-tetrachloroethane in a phosphate buffer containing 50% (v/v) ethanol to female NMRI mice at 1.2 to 20 g/kg resulted in the presence of the parent compound at about 43% (21 to 62%) of the administered dose in the expired air at the end of a 3-day collection period. The principal urinary and fecal metabolites after 3 days were trichloroethanol and trichloroacetic acid at about 31% (17 to 49%) and 4% (1 to 7%) of dose administered, respectively.

- Thompson et al. (1984), Town and Liebman (1984) and Salmon et al. (1985) have found evidence for the cytochrome P450-dependent reductive dechlorination of 1,1,1,2-tetrachloroethane in microsomal preparations.

D. Excretion

- Yllner (1971) determined that the amount of unchanged parent compound in the expired air of female NMRI mice receiving 1.2 to 2.0 g/kg subcutaneous 1,1,1,2-tetrachloroethane averaged 43% (21 to 62%) of the administered dose, within 3 days after dosing. Combined urinary and fecal excretion of trichloroethanol and trichloroacetic acid accounted for about 31% (17 to 49%) and about 4% (1 to 7%) of the administered dose, respectively. About 78% (72 to 82%) of the dose had been excreted within 3 days.

- Ikeda and Ohtsuji (1972) reported the excretion of urinary metabolites (trichloro compounds) equivalent to 199 mg/kg by rats exposed to 1,1,1,2-tetrachloroethane at 200 ppm (1,373 mg/m³) for 8 hours and the excretion of urinary metabolites equivalent to 114.2 mg/kg by rats injected intraperitoneally with 2.78 mmol/kg (467 mg/kg) 1,1,1,2-tetrachloroethane. The collection period was 48 hours. Materials excreted through exhalation were not measured.

- Mitoma et al. (1985) administered a single oral dose of ^{14}C-1,1,1,2-tetrachloroethane in corn oil to male Osborne-Mendel rats (200 mg/kg) and B6C3F$_1$ mice (400 mg/kg) following 4 weeks of oral administration (5 days/week) of the unlabeled compound. Expired air and excreta collected for 48 hours after administration of the ^{14}C-labeled compound were analyzed. As a percentage of the radioactive dose, 34.14% (rat) and 5.89% (mouse) were expired unchanged, 1.47% (rat) and 2.08% (mouse) were expired as CO_2, 60.09% (rat) and 77.21% (mouse) were recovered from excreta, and 3.20% (rat) and 5.04% (mouse) were recovered from carcasses.

III. Health Effects

A. Humans

1. Short-term Exposure

- Prolonged exposure to tetrachloroethane fumes (isomer not specified) produced weakness and nausea associated with acute hepatic damage (Norman et al., 1981). Exposure levels and duration were not reported.

2. Long-term Exposure

- Norman et al. (1981) retrospectively compared mortality records of men exposed to tetrachloroethane (isomer not specified), in chemical processing plants in which clothing was impregnated for protection against mustard gas exposure during World War II, to those not exposed to tetrachloroethane (i.e., working in chemical processing plants using a water-based solvent instead of tetrachloroethane). Exposure duration ranged from 5 weeks to 1 year. Exposure levels were not quantitated. Although not statistically significant, a slight increase in relative risk (RR) of death due to genital (RR = 4.56) and lymphatic (RR = 5.19) cancers and leukemias (RR = 1.77) was noted in 1,099 tetrachloroethane-exposed workers, compared with 1,319 unexposed workers in the same companies. Overall cancer mortality for tetrachloroethane-exposed workers was 1.26 times that of unexposed workers. Interpretation of these results is difficult, however, since workers were also exposed to the impregnite, N-dichloro-hexachloro-diphenyl-urea and dry-cleaning solvents.

B. Animals

1. Short-term Exposure

- The NTP (1983) administered a single dose of 1,1,1,2-tetrachloroethane by gavage in corn oil to groups of 5 male and 5 female B6C3F$_1$ mice and 5 male and 5 female F344/N rats at 0, 10, 100, 500, 1,000 or 5,000 mg/kg, and all animals were observed for 14 days after the single dose. The high dose caused the deaths of all treated animals. At the 1,000 mg/kg dose, 1/5 male rats and 3/5 female rats, but no mice, died. No rats or mice were killed by lower doses, and no compound-related gross pathologic effects were observed in any animal.

- Truhaut et al. (1974) found oral LD$_{50}$ values of 1,500, 670 and 780 mg/kg in male Swiss-Webster mice, male Wistar rats and female Wistar rats, respectively. At doses of 1,000 to 2,500 mg/kg, mice developed microvacuolizations and severe centrilobular necrosis. In male rats receiving 1,000 mg/kg, the liver was granular with centrilobular cytoplasmic swellings; females given 800 mg/kg had severe microvacuolar steatosis, and lungs were congested and the spleens contained numerous macrophages and hemosiderin deposits. Doses of 300 mg/kg (male rats) and 400 mg/kg (female rats) each resulted in 0/10 deaths after 14 days. Histopathological alterations at these levels, if any, were not discussed. Cutaneous LD$_{50}$ was 20 g/kg. Its acute toxicity by inhalation, for an exposure of 4 hours, was 2,100, 2,500 and 2,800 mL/m^3 for male and female rats, and male rabbits, respectively.

- Truhaut et al. (1973) reported elevations in serum glutamic pyruvic transaminase (SGPT or ALT), serum glutamic oxalacetic transaminase (SGOT or AST), creatinine phosphokinase (CPK), lactate dehydrogenase (LDH) and LDH isozyme activities, measured 24 hours after exposure, and noted aberrant electrocardiographic O waves in seven rabbits given 500 mg/kg 1,1,1,2-tetrachloroethane by gavage. Histopathology revealed microvacuoles and centrilobular necrosis of the liver. Blood cholesterol and total lipid levels were slightly elevated up to 96 hours after dosing. Truhaut et al. (1974) administered the same oral dose to separate groups of guinea pigs and rabbits, and sacrificed animals 1 day to 1 month thereafter. After 1 day (guinea pigs) and 3 days (rabbits), severe centrilobular necrosis developed in the livers of most animals. Heart tissue was characterized by edematous dissociation of myocardial fibrils.

- Truhaut et al. (1975) observed hepatic steatosis accompanied by an accumulation of triglycerides and decreases in hepatic LDH, malate dehydrogenase (MDH) and glutamic pyruvic transaminase (GPT), in female Wistar rats (number not specified) given 300 mg/kg 1,1,1,2-tetrachloroethane by gavage in olive oil 5 days/week for 2 weeks. These changes were not seen in the livers of male rats at the same dosages when compared with the male or female vehicle controls.

- Truhaut et al. (1974) administered 400 mg/kg 1,1,1,2-tetrachloroethane in olive oil, by gavage, to 10 Wistar rats/sex, 5 days/week for 2 weeks. Seven male rabbits received 500 mg/kg for 2 days. Dosing was fatal to one male rat, two female rats and four rabbits. Surviving female rats displayed a significant weight loss.

- The NTP (1983) treated five F344/N rats and five B6C3F₁ mice/sex with 0, 10, 50, 100, 500 or 1,000 mg/kg/day 1,1,1,2-tetrachloroethane by gavage in corn oil for 14 consecutive days. Dosing was fatal to 1/5 female rats at 500 mg/kg, and to 3/5 male and 1/5 female rats at 1,000 mg/kg. Inhibition of body weight gain was observed only in high-dose rats, and no compound-related effects were observed at necropsy. At the high dose, 1/5 males and 2/5 females died, although this treatment had no effect on body weights or gross pathology.

2. Dermal/Ocular Effects

- Truhaut et al. (1974) showed that 0.5 mL of 1,1,1,2-tetrachloroethane, applied to the shaved skin of rabbits produced a slowly forming and reversible erythema. Histological examination revealed epidermal atrophy, ulcerations and small areas of dermal necrosis. Ocular treatment of rabbits with 0.1 mL 1,1,1,2-tetrachloroethane led to a marked corneal opacity and conjunctivitis.

3. Long-term Exposure

- In a 13-week study, the NTP (1983) administered 0, 5, 10, 50, 100 or 500 mg/kg 1,1,1,2-tetrachloroethane by gavage in corn oil 5 days/week to groups of 10 male and 10 female F344/N rats and 10 male and 10 female B6C3F₁ mice. Animals were checked daily for mortality and morbidity, clinical examinations and body weight measurements were made weekly, necropsies were performed on all animals not autolyzed or cannibalized and selected tissue sites from animals in the control and high-dose conditions were examined histologically. Both sexes of rats given the high dose had a 7 to 8% inhibition in body weight gain. One high-dose rat died on each of weeks 10 (male) and 11 (female). One high-dose male mouse died during week 3 and one female rat administered 100 mg/kg died at week 2. Female rats administered 500 mg/kg also exhibited loss of equilibrium. Body weights were unaffected by treatment in mice. Neither species had compound-related effects on histological parameters.

- Truhaut et al. (1974) administered 300 mg/kg 1,1,1,2-tetrachloroethane by gavage, 5 days/week for 10 months, to groups of 10 male and 10 female Wistar rats. Control rats received olive oil vehicle only. This dosing regime was fatal to 10% of the controls, 25% of the treated males and 40% of the treated females. Hepatic damage in treated rats, although variable, was quite severe, particularly in females.

- The NTP (1983) conducted a 103-week study in F344/N rats exposed to 0, 125 or 250 mg/kg 1,1,1,2-tetrachloroethane in corn oil by gavage, 5 days/week. Initially, 50 males and 50 females were in each dose group. Rats were observed twice daily for clinical signs, weighed monthly, and, upon death or when terminally sacrificed, were necropsied and microscopically examined. Although there were no differences in body weights due to treatment, the high-dose rats were weak and uncoordinated beginning on week 44, and the survival rate of high-dose males was significantly less ($p < 0.001$) than that of controls. Interpretation of mortality data was complicated by the accidental killing of 27 males and 15 females by either gavage error or heat stress. Mineralization (probably calcification) of the kidneys was dose-related in males, and hepatic clear cell changes and pulmonary alveolar emphysema was dose-related in females. The investigators considered the latter effect to be a result of the intubation process. A LOAEL of 125 mg/kg has been identified from this study.

- The NTP (1983) also intubated 50 male and 50 female B6C3F$_1$ mice/group with 0, 250 or 500 mg/kg 1,1,1,2-tetrachloroethane in corn oil, 5 times/week for 103 weeks. End points were the same as those observed in the rat study. The high-dose mice began to show an inhibition in body weight gain at week 20, and by week 44 were weak, uncoordinated and had breathing difficulties. By week 65, all high-dose mice were either dead or moribund; the latter were subsequently sacrificed. At 500 mg/kg, there were marked elevations in incidences of hepatic inflammation, focal and diffuse necrosis, fatty metamorphosis and cytomegaly in both sexes of mice. A decrease in uterine cystic hyperplasia, probably due to decreased longevity (NTP, 1983), was also noted in high-dose females.

In the low-dose female mice, survival was significantly (p = 0.039) reduced relative to control values.

4. Reproductive Effects

- As part of a long-term experiment described earlier, Truhaut et al. (1974) found that the reproductive functions of the adult rats were not impaired.

5. Developmental Effects

- As part of the long-term experiment described earlier, Truhaut et al. (1974) found that all pups were dead within 48 hours of birth with severe microvacuolar steatosis.

6. Mutagenicity

- Simmon et al. (1977) found 1,1,1,2-tetrachloroethane not to be genotoxic to *Salmonella typhimurium* TA1535, TA1537, TA1538, TA98 or TA100, with or without metabolic activation.

- 1,1,1,2-Tetrachloroethane was negative in the rat liver foci initiation/promotion assay (Story et al., 1986).

7. Carcinogenicity

- The NTP (1983) conducted a bioassay in which 50 F344/N rats/sex received 0, 125 or 250 mg/kg 1,1,1,2-tetrachloroethane in corn oil by gavage 5 days/week for 103 weeks. Survival of male rats was significantly reduced, relative to control rats, and 27 male and 15 female rats were accidentally killed by either heat stress or gavage error. The numbers of low-dose males and high-dose females at the termination of the study were not statistically significantly lower than the numbers of respective controls. By the life table trend test, there was a statistically significant (p = 0.044) increase in combined hepatic neoplastic nodules/carcinomas in male rats. Mammary fibroadenoma incidence in females was significantly (p < 0.05) increased in low-dose female rats by the Fisher exact test, although there were no differences between high-dose females and controls. The NTP (1983) did not consider the available data sufficiently persuasive to warrant classification of 1,1,1,2-tetrachloroethane as a carcinogen in F344/N rats, under the conditions of this study.

- The NTP (1983) found more definitive evidence for oncogenicity in 50 B6C3F$_1$ mice/sex/group exposed to 0, 250 or 500 mg/kg 1,1,1,2-tetrachloroethane by gavage in corn oil 5 days/week for 103 weeks. Most neoplastic lesions were restricted to the liver. In both male and female mice, there were significant, positive, dose-related trends in the incidences of hepatocellular adenomas and carcinomas. Both the NTP (1983) and the peer review committee cited these results as evidence of a causal association between 1,1,1,2-tetrachloroethane administration and liver tumors in mice.

IV. Quantification of Toxicological Effects

A. One-day Health Advisory

The Ten-day HA for a 10-kg child (2 mg/L) is adopted as the One-day HA for a 10-kg child. Several single dose oral studies were available but were inadequate for quantitative risk assessment. In an NTP (1983) gavage experiment in rats and mice, only mortality and gross lesions were evaluated. Gross lesions were not observed in rats or mice killed with single large (5,000 mg/kg) doses of 1,1,1,2-tetrachloroethane. Truhaut et al. (1973) observed effects on the livers and hearts of guinea pigs and rabbits intubated with 500 mg/kg, but lower doses were not used and a NOAEL was not determined. Truhaut et al. (1974) observed no mortality in male rats at 300 mg/kg or in females at 400 mg/kg, but other parameters of toxicity were not discussed.

B. Ten-day Health Advisory

The study by Truhaut et al. (1975) has been selected to serve as the basis for the 10-kg child Ten-day HA because it defines a LOAEL in female rats exposed by a relevant route. Although the clinical significance of decreases in LDH, MDH and GPT activities is unclear, triglyceride accumulation and hepatic steatosis can be considered to be adverse effects. These effects were not observed in the livers of treated males when compared with the vehicle controls.

The 2-week study by Truhaut et al. (1974) in rats and rabbits defined only frank effects, including increased mortality and weight loss, at a Time-weighted Average (TWA) oral dose of 286 mg/kg/day (400 mg/kg/day x 5 days exposure/7 days). Based upon mortality, body weight and gross necropsy data, an apparent NOAEL of 100 mg/kg/day was defined in the 2-week NTP (1983) bioassay. However, microscopic analysis of tissue from treated animals was not provided.

The Ten-day HA for the 10-kg child is calculated as follows:

$$\text{Ten-day HA} = \frac{(300 \text{ mg/kg}) (5/7) (10 \text{ kg})}{(1,000) (1 \text{ L/day})} = 2.1 \text{ mg/L [rounded to } 2 \text{ mg/L (2,000 } \mu\text{g/L)]}$$

where:

$$5/7 = \text{conversion from 5 to 7 days of exposure.}$$

C. Longer-term Health Advisory

The Longer-term HAs for a child and adult are based on the LOAEL of 125 mg/kg administered 5 days/week to rats in a 2-year gavage experiment (NTP, 1983). Two longer-term gavage studies were reviewed but were inadequate for quantitative risk assessment. In a 13-week study, a reduced rate of body weight gain and mortality of 2/20 was observed in rats at 500 mg/kg. Treatments were performed 5 days/week. No adverse effects were noted in rats at ≤ 100 mg/kg or in mice at 500 mg/kg, the highest dose tested. Histopathological examination, performed only on controls and high-dose (500 mg/kg) rats and mice, revealed no lesions (NTP, 1983). Truhaut et al. (1974) observed death and liver damage in rats intubated with 300 mg/kg, 5 days/week for 10 months. A NOAEL for kidney and liver damage is not determined in these experiments and it appears that increasing duration of exposure results in the appearance of more severe effects.

The Longer-term HA for a 10-kg child is calculated as follows:

$$\text{Longer-term HA} = \frac{(125 \text{ mg/kg}) \ (5/7) \ (10 \text{ kg})}{(1,000) \ (1 \text{ L/day})} = 0.9 \text{ mg/L } (900 \ \mu g/L)$$

where:

$$5/7 = \text{conversion from 5 to 7 days of exposure.}$$

The Longer-term HA for a 70-kg adult is calculated as follows:

$$\text{Longer-term HA} = \frac{(125 \text{ mg/kg}) \ (5/7) \ (70 \text{ kg})}{(1,000) \ (2 \text{ L/day})} = 3.1 \text{ mg/L [rounded to } 3 \text{ mg/L } (3,000 \ \mu g/L)]$$

where:

$$5/7 = \text{conversion from 5 to 7 days of exposure.}$$

D. Lifetime Health Advisory

The DWEL for a 70-kg adult is based on the LOAEL of 125 mg/kg 1,1,1,2-tetrachloroethane administered in corn oil 5 days/week to rats in a 2-year gavage experiment (NTP, 1983).

Step 1: Determination of the Reference Dose (RfD)

$$\text{RfD} = \frac{(125 \text{ mg/kg}) \ (5/7)}{(1,000) \ (3)} = 0.0298 \text{ mg/kg/day (rounded to } 0.03 \text{ mg/kg/day)}$$

where:

5/7 = conversion from 5 to 7 days of exposure.

3 = additional uncertainty factor for lack of adequate supporting reproductive and chronic toxicity studies.

Step 2: Determination of the Drinking Water Equivalent Level (DWEL)

$$\text{DWEL} = \frac{(0.0298 \text{ mg/kg/day}) (70 \text{ kg})}{(2 \text{ L/day})} = 1.04 \text{ mg/L [rounded to}$$
$$1 \text{ mg/L } (1,000 \text{ } \mu g/L)]$$

Step 3: Determination of the Lifetime Health Advisory

$$\text{Lifetime HA} = \frac{(1.04 \text{ mg/L}) (20\%)}{(3)} = 0.0694 \text{ mg/L [rounded to}$$
$$0.07 \text{ mg/L } (70 \text{ } \mu g/L)]$$

where:

3 = additional uncertainty factor per ODW policy to account for possible carcinogenicity. For this chemical, the quantitative cancer risk assessment indicates that an additional uncertainty factor of 3 is necessary to account for possible cancer risk.

E. Evaluation of Carcinogenic Potential

- The U.S. EPA (IRIS, 1989) derived an upper limit for carcinogenic potency factor (q_1*) of 2.6×10^{-2} (mg/kg/day) using the incidences of combined hepatocellular adenoma and carcinoma in female mice from the NTP (1983) bioassay. Using this q_1*, the 95% upper limit lifetime doses associated with 10^{-4}, 10^{-5} and 10^{-6} risk levels are calculated to be 4, 0.4 and 0.04 $\mu g/kg/day$, respectively. Assuming an average water consumption of 2 L/day, and a human reference body weight of 70 kg, these risk levels correspond to water concentrations of 100, 10 and 1 $\mu g/L$, respectively.

- In the chronic rat cancer experiment (NTP, 1983), there were increased incidences of mammary fibroadenomas in low-dose females and a slight dose-related trend in combined hepatocellular neoplastic nodules/carcinomas in the livers of male rats. Neither of these findings is sufficient to implicate 1,1,1,2-tetrachloroethane as a carcinogen in this strain. In the mouse study, the NTP (1983) noted dose-related increases in hepatocellular adenomas and carcinomas in both sexes of mice, implicating the chemical as a carcinogen in this species.

- Applying the criteria described in the U.S. EPA's Guidelines for Carcinogen Risk Assessment (U.S. EPA, 1986a), 1,1,1,2-tetrachloroethane has been classified in Group C: possible human carcinogen. This category is for agents with limited evidence of carcinogenicity in animals in the absence of human data.

V. Other Criteria, Guidance and Standards

- Other criteria, guidelines or standards regarding or regulating 1,1,1,2-tetrachloroethane could not be located in the available literature.

VI. Analytical Methods

- Analysis of 1,1,1,2-tetrachloroethane is by a purge-and-trap gas chromatographic procedure used for the determination of volatile organohalides in drinking water (U.S. EPA, 1985a). This method calls for the bubbling of an inert gas through the sample and trapping volatile compounds on an adsorbent material. The adsorbent material is heated to drive off the compounds onto a gas chromatographic column. The gas chromatograph is temperature programmed to separate the method analytes, which are then detected by a halogen-specific detector. Confirmatory analysis is by mass spectrometry (U.S. EPA, 1985b). The detection limit has not been determined for either method.

VII. Treatment Technologies

- Available data indicate that air stripping and possibly granular activated carbon (GAC) adsorption will remove 1,1,1,2-tetrachloroethane from contaminated water.

- Beaudet et al. (1981) conducted a 3-year study to remove priority pollutants from wastewater by GAC. Tetrachloroethane was present in concentrations ranging from 0.01 mg/L to 130 mg/L with an average concentration of 20 mg/L. A pilot-plant GAC system, containing three columns packed with Filtrasorb 300 granular carbon, was operated at 20-, 40-, and 60-minute cumulative empty bed contact times (EBCT). Tetrachloroethane was effectively removed by column 1, while columns 2 and 3 did not reach breakthrough (0.1 mg/L) on this compound prior to the end of the pilot operations. No indication was given with respect to which tetrachloroethane isomer(s) were present or removed. No other operating data are provided.

- Kincannon et al. (1983) investigated optimum removal mechanisms for various priority pollutants from wastewater. The authors report 94.5% removal of 1,1,2,2-tetrachloroethane through a complete mix, continuous flow, activated sludge system. They attribute all of this removal to air stripping.

- U.S. EPA (1986b) estimated the feasibility of removing 1,1,1,2- tetrachloroethane from water by air stripping, employing the engineering design procedure and cost model presented at the 1983 National ASCE Conference on Environmental Engineering. Based on chemical and physical properties, and assumed operating conditions, a 99% removal efficiency of 1,1,1,2-tetrachloroethane was reported by a column with a diameter of 5.5 ft and packed with 20 ft of 1-inch plastic saddles. The air-to-water ratio required to achieve this degree of removal effectiveness is 6.2.

VIII. References

Archer, W.L. 1979. Chlorocarbons-hydrocarbons (other). In: Grayson, M. and D. Eckroth, eds., Kirk-Othmer Encyclopedia of Chemical Technology, Vol. 5, 34th ed., New York, NY: John Wiley and Sons, Inc. p. 734.

Beaudet, B.A., L.J. Bilello, E.M. Kellar, J.M. Allan and R.J. Turner. 1981. Removal of specific organics from wastewater by activated carbon adsorption: evaluation of a rapid method for determining carbon usage rates. Proc. Ind. Waste Conf. 35:381–391.

Dilling, W.L. 1977. Interphase transfer processes. II. Evaporation rates of chloromethanes, ethanes, ethylenes, propanes and propylenes from dilute aqueous solutions. Comparisons with theoretical predictions. Environ. Sci. Technol. 11:405–409.

HSDB. 1988. Hazardous Substance Data Bank. Online. Bethesda, MD: National Library of Medicine.

Ikeda, M. and H. Ohtsuji. 1972. Comparative study of the excretion of Fujiwara reaction-positive substances in urine of humans and rodents given trichloro-or tetrachloro-derivatives of ethane and ethylene. Br. J. Ind. Med. 29:99–104.

IRIS. 1989. Integrated Risk Information System. Online. Washington, DC: U.S. Environmental Protection Agency.

Keith, L.H., A.W. Garrison, F.R. Allen et al. 1976. Identification of organic compounds in drinking water from thirteen U.S. cities. In: L.H. Keith, ed., Identification and Analysis of Organic Pollutants in Water. Ann Arbor, MI: Ann Arbor Science. pp. 329–373.

Kincannon, D.F., E.L. Stover, V. Nichols and D. Medley. 1983. Removal mechanisms for toxic priority pollutants. J. Water Pollut. Control Fed. 55(2):157–163.

Mitoma, C., T. Steeger, S.E. Jackson, K.P. Wheeler et al. 1985. Metabolic disposition study of chlorinated hydrocarbons in rats and mice. Drug Chem. Tox. 8(3):183–194.

Norman, J.E., C.D. Robinson and J.F. Fraument, Jr. 1981. The mortality experience of Army World War II chemical processing companies. J. Occup. Med. 23:818–822.

NTP. 1983. National Toxicology Program. Carcinogenesis studies of 1,1,1,2-tetrachloroethane in F344/N rats and B6C3F$_1$ mice (gavage studies). NTP Carcinogenesis. Tech. Rep. Ser. No. 237. p. 146. (Also published as NIH Publ. No. 83–1793.)

Pearson, C.R. and G. McConnell. 1975. Chlorinated C$_1$ and C$_2$ hydrocarbons in the marine environment. Proc. Royal Soc. London. Ser. B. 189:305.

Salmon, A.G., J.A. Nash, C.M. Walklin and R.B. Freedman. 1985. Dechlorination of halocarbons by microsomes and vesicular reconstituted cytochrome P-450 systems under reductive conditions. Br. J. Ind. Med. 42:305–311.

Scheiman, M.A., R.A. Saunders and F.E. Saalfield. 1974. Organic contaminants in the District of Columbia drinking water supply. Biomed. Mass. Spectrom. 1(4):209–211.

Simmon, V.F., K. Kauhanen and R.G. Tardiff. 1977. Mutagenic activity of chemicals identified in drinking water. 2nd Int. Cong. Environ. Mutagens, Edinburgh, Scotland, July, 1977. Dev. Toxicol. Environ. Sci. 2:249–258.

Singh, H.B., L.J. Salas, A.J. Smith and J. Shigeishi. 1981. Measurements of some potentially hazardous organic chemicals in urban environments. Atmos. Environ. 15:601–612.

Story, D.L., E.F. Meierhenry, C.A. Tyson and H.A. Milman. 1986. Differences in rat liver enzyme-altered foci produced by chlorinated aliphatics and phenobarbital. Toxicol. Ind. Health. 2(4):351–362.

Tabak, H.H., S.A. Quave, C.I. Mashni and E.F. Barth. 1981. Biodegradability studies with organic priority pollutant compounds. J. Water Pollut. Control Fed. 53:1503–1518.

Thompson, J.A., B. Ho and S.L. Mastovich. 1984. Reductive metabolism of 1,1,1,2-tetrachloroethane and related chloroethanes by rat liver microsomes. Chem. Biol. Interact. 51(3):321–333.

Town, C. and K.C. Liebman. 1984. The *in vitro* dechlorination of some polychlorinated ethanes. Drug Metab. Dispos. 12:4–8.

Truhaut, R. 1972. Metabolic transformations of 1,1,1,2-tetrachloroethane in animals (rats, rabbit). Chem. Anal. (Warsaw) 17(4):1075–1078. (In French with English summary.)

Truhaut, R., N.P. Lich, N.T. Le Quang Thuan and H. Dutertre-Catella. 1973.

Serum enzymic activities and biochemical blood components in subacute 1,1,1,2-tetrachloroethane poisoning in the rabbit. J. Eur. Toxicol. 6(2):81–84. (In French with English summary.)

Truhaut, R., N.P. Lich, H. Dutertre-Catella, G. Molas and V.N. Huyen. 1974. Toxicological study of 1,1,1,2-tetrachloroethane. Arch. Mal. Prof. Med. Trav. Secur. Soc. 35(6):593–608. (In French with English summary.)

Truhaut, R., M. Thevenin, J.M. Warner, J.R. Claude and N.P. Lich. 1975. Preliminary biochemical study of the hepatotoxicity of 1,1,1,2-tetrachloroethane in the Wistar rat. Effect of sex. Eur. J. Toxicol. Environ. Hyg. 8(3):175–179. (In French with English summary.)

U.S. EPA. 1978. U.S. Environmental Protection Agency. Quantification of chlorinated hydrocarbons in previously collected air samples. Research Triangle Park, NC. EPA 450/3-78-112. NTIS PB 289-804.

U.S. EPA. 1979. U.S. Environmental Protection Agency. Water-related environmental fate of 129 priority pollutants, Vol. II. Office of Water Planning and Standards, Office of Water and Waste Management, Washington, DC. EPA 400/4-79-029b. NTIS PB 80-204381.

U.S. EPA. 1985a. U.S. Environmental Protection Agency. U.S. EPA Method 502.1–Volatile halogenated organic compounds in water by purge and trap gas chromatography. Environmental Monitoring and Support Laboratory, Cincinnati, OH 45268, June 1985 (Revised November 1985).

U.S. EPA. 1985b. U.S. Environmental Protection Agency. U.S. EPA Method 524.1–Volatile organic compounds in water by purge and trap gas chromatography/mass spectrometry. Environmental Monitoring and Support Laboratory, Cincinnati, OH 45268, June 1985 (Revised November 1985).

U.S. EPA. 1986a. U.S. Environmental Protection Agency. Guidelines for carcinogen risk assessment. *Fed. Reg.* 51(185):33992–34003. September 24.

U.S. EPA. 1986b. U.S. Environmental Protection Agency. Economic evaluation of 1,1,1,2-tetrachloroethane removal from water packed column air stripping. Prepared by the Office of Drinking Water for Health Advisory Treatment Summaries.

Yllner, S. 1971. Metabolism of 1,1,1,2-tetrachloroethane in the mouse. Acta. Pharmacol. Toxicol. 29(5–6):471–480.

Chloromethane

I. General Information and Properties

A. CAS No. 74–87–3

B. Structural Formula

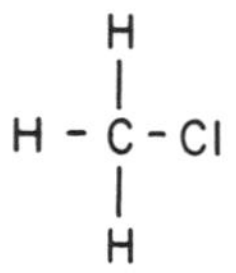

$$
\begin{array}{c}
\text{H} \\
| \\
\text{H} - \text{C} - \text{Cl} \\
| \\
\text{H}
\end{array}
$$

Chloromethane

C. Synonyms

* Methyl chloride, monochloromethane.

D. Uses

* Chloromethane is used in the production of other chemical products. As such, it has the following use pattern (in percentages of total use in the United States) (CMR, 1983):

Silicones	60%
Tetramethyl lead	15%
Methyl cellulose	6%
Herbicides	4%
Quarternary amines	4%
Butyl rubber	3%
Miscellaneous	3%
Exports	5%

E. Properties (Ahlstrom and Steele, 1979; Mackay and Shiu, 1981; Hansch and Leo, 1985)

Chemical Formula	CH_3Cl
Molecular Weight	50.49
Physical State (at 25°C)	Colorless gas
Boiling Point (760 mm Hg)	−23.73°C
Melting Point	−97.7°C
Density	−
Vapor Pressure (25°C)	4,275 mm Hg
Specific Gravity (20°C)	0.920 (liquid)
Water Solubility (25°C)	5,350 mg/L
	4,800 mg/L
Log Octanol/Water Partition Coefficient	0.91
Taste Threshold (water)	−
Odor Threshold (water)	−
Odor Threshold (air)	21 mg/m³
Conversion Factor (in air)	1 ppm = 2.06 mg/m³
	1 mg/m³ = 0.48 ppm

F. Occurrence

- Chloromethane has been detected in drinking water in Miami, FL; Ottumwa, IA; Philadelphia, PA; and Cincinnati, OH (U.S. EPA, 1977). It has been detected at levels < 5 ppb in about 3% of Canadian drinking water samples from treatment plants (Otson et al., 1982). Page (1981) monitored 1,058 ground-water sites and detected chloromethane in only three samples with a maximum concentration of 6.0 ppb. Burmaster (1982) reported a concentration of 44 ppb in a well in Massachusetts. Greenberg et al. (1982) monitored 408 well water samples from New Jersey and qualitatively detected chloromethane in < 10 samples.

- Singh et al. (1983) reported an average concentration of 11.5 ng/L (11.5 ppt) chloromethane in the surface water of the Pacific Ocean. Chloromethane was detected in 24/605 surface waters examined in New Jersey, where the highest concentration reported was 222.4 ppb, although the median concentration was below the detection limit (Page, 1981).

- The average atmospheric concentration of chloromethane in seven U.S. urban areas was reported to be 665 to 995 ppt (1.37 to 1.97 µg/m³) (Singh et al., 1982).

G. Environmental Fate

- At 25°C, the hydrolysis half-life of chloromethane in water was estimated to be 0.93 years with the hydrolysis rate independent of pH at pH < 10 (Mabey and Mill, 1978). Callahan et al. (1979) estimated the half-life to be 2.5 years.

- Based on its low absorption cross-section in the gas phase at wavelengths > 290 nm (Crutzen et al., 1978), significant direct photolysis of chloromethane in water does not seem likely.

- The volatilization half-life of chloromethane from a river 1 meter (m) deep, flowing with a current speed of 1 m/sec and a windspeed of 3 m/sec was calculated to be 2.4 hours at 20°C (Lyman et al., 1982). Therefore, volatilization is the most important removal mechanism from aquatic media.

- U.S. EPA (1986b) estimated that the biodegradation half-life of chloromethane in natural water is 19 days. Therefore, biodegradation is not a significant aquatic degradation process.

II. Pharmacokinetics

A. Absorption

- Data regarding the gastrointestinal absorption of chloromethane could not be located in the available literature.

- According to ACGIH (1980), the chemical "can be absorbed through the skin," but quantitative data were not provided.

- Nolan et al. (1985) exposed six male volunteers to 10 or 50 ppm (21 or 103 mg/m^3) chloromethane for 6 hours. Blood and expired air concentrations of chloromethane reached equilibrium during the first hour. The expired air contained 30 to 70% of the concentration of chloromethane in inhaled air. Absorption rates of 1.4 to 3.7 mg/min/kg were calculated using a two-compartment pharmacokinetic model.

- Alveolar air levels reached equilibrium within 1 hour when humans were exposed via inhalation to 100 or 200 ppm (207 or 413 mg/m^3) chloromethane (Putz-Anderson et al., 1981a). The alveolar levels were 36 and 63 ppm (74 and 130 mg/m^3), respectively.

- Landry et al. (1983) measured chloromethane uptake in male F344 rats after 6 hours of exposure via inhalation to 50 or 1,000 ppm (103 or 2,065 mg/m^3) with calculated uptakes of 0.028 and 0.540 mg/min/kg.

B. Distribution

- After 59 to 110 mg chloromethane in a propylene glycol vehicle was administered intravenously to dogs, blood levels declined rapidly (90% lost immediately and 98% lost in 61 hours) (Sperling et al., 1950). Immediately after injection, chloromethane was detected in the brain, heart, liver, stomach, spleen and blood at levels generally < 0.02 mg/g tissue.

C. Metabolism

- In an *in vitro* study using tissue homogenates from rats, chloromethane was conjugated with sulfhydryl groups in erythrocytes, liver, brain and kidneys to form S-methylcysteine and S-methylglutathione. Hydrolysis of S-methylglutathione to S-methylcysteine occurred in the red blood cells, brain and kidneys, but not in the liver (Redford-Ellis and Gowenlock, 1971).

- Concentration-dependent depletion of nonprotein sulfhydryl groups (NPSH) (primarily glutathione) occurred in the liver, kidney and brain of rats and mice after exposure to chloromethane via inhalation at concentrations ranging from 100 to 2,500 ppm (207 to 5,160 mg/m^3) for 6 hours (Kornbrust and Bus, 1984). Although the mouse liver NPSH levels were significantly decreased (55% of controls) after exposure to 100 ppm of chloromethane for 6 hours, no effect on the rat liver NPSH levels was observed at this level of exposure. However, after an exposure to 500 ppm for 6 hours, rats also had decreased (41% of controls) liver NPSH levels (Dodd et al., 1982).

- After conjugation to glutathione, chloromethane is metabolized (probably through the intermediates formaldehyde and formate) to a species that enters the one-carbon cytoplasmic pool. Formate was detected in rats exposed to chloromethane only after inhibition of formate metabolism by nitrous oxide. The amount of formate identified was significantly increased in rats exposed to nitrous oxide and chloromethane in comparison with rats treated with nitrous oxide alone (Kornbrust and Bus, 1982).

- Kornbrust and Bus (1983) concluded that methanethiol, derived from S-methylcysteine through the intermediates methylthiopyruvic acid and methylthioacetic acid, might be responsible for the toxic effects of chloromethane. Alternatively, toxic effects may be due to formaldehyde accumulation of GSH depletion.

- Nolan et al. (1985) reported that there are two discrete groups of humans in regard to the rate of chloromethane metabolism: one rapid (capable of clearing 3,460 mL of blood/minute), the other slow (capable of clearing 795 mL of blood/minute).

- Elimination of chloromethane from the blood follows second-order kinetics and is rapid once inhalation exposure ceases with a half-life of 15 minutes in the rat, 50 minutes in the dog and in humans that are rapid metabolizers, and 90 minutes in humans that are slow metabolizers (Landry et al., 1983; Nolan et al., 1985).

D. Excretion

- After intravenous injection of chloromethane in dogs, only 5% of the total administered dose was recovered unchanged from the expired air, and renal and biliary excretion of chloromethane was almost negligible (Sperling et al., 1950).

- After rats inhaled ^{14}C-chloromethane (1,500 ppm or 3,100 mg/m³), 45% of the inhaled dose of radioactivity was recovered in 2 to 24 hours postexposure from the expired air as carbon dioxide, and an additional 6.6% as chloromethane (Kornbrust et al., 1982).

- Bus (1978) reported that 63.9, 32 and 3.9% of the total excreted radioactivity were recovered from the air, urine and feces, respectively, of rats exposed to 1,500 ppm of ^{14}C-chloromethane for 6 hours; however, the total percentage of the administered dose that was recovered through all three routes of excretion was not reported. Excretion was essentially complete by 18 hours.

III. Health Effects

A. Humans

1. Short-term Exposure

- Chloromethane is considered to be a relatively severe narcotic in humans. Acute human inhalation exposure to chloromethane primarily affects the central nervous system, causing headache, giddiness, vision disturbances, sleepiness, unconsciousness, and in severe cases, convulsions, opisthotonus and death (Browning, 1965; Scharnweber et al., 1974; Repko and Lasley, 1979; Repko, 1981; Torkelson and Rowe, 1981). In addition to central nervous system effects, the gastrointestinal tract may be another site of significant effect. [Involvement of liver, kidney and heart appears to be unusual (Repko and Lasley, 1979)]. The symptoms may be delayed in onset and may increase in severity for 24 to 48 hours after exposure.

- No effect on hand-eye coordination, mental alertness or time discrimination was observed in a group of 12 humans exposed to 200 ppm (413 mg/m³) of chloromethane for 3.5 hours (Putz-Anderson et al., 1981b).

- A slight performance impairment was observed in the hand-eye coordination, mental alertness and time discrimination of a group of 12 university students exposed to 200 ppm (413 mg/m³) of chloromethane via inhalation for 3 hours. The study included male and female students with a mean age of 22 years (Putz-Anderson et al., 1981a). [There were large individual differences in chloromethane concentration of both alveolar air and blood.]

2. Long-term Exposure

- Repko et al. (1976) studied the effect of occupational chloromethane exposure via inhalation on 122 workers (exposed from 1 to 26 years) in manufacturing plants utilizing chloromethane. The control group consisted of 49 nonexposed workers from the same industry. The concentration of chloromethane in the ambient air of the exposed workers ranged from 7.4 to 70 ppm, with a mean concentration of 34 ppm (70 mg/m³). A battery of behavioral, psychological and neurological tests, and an electroencephalogram (EEG) was done on each worker. An increase in ambient air concentration and an increase in urine acidity (both indicators of chloromethane exposure) were correlated with a poorer performance on the behavioral tasks. There was no relationship between exposure and psychological or personality effects. Likewise, there was no effect on neurological tests or EEG records. However, chloromethane exposure was correlated with an adverse effect on the performance of cognitive time-sharing tasks and a significant increase in the magnitude of finger tremors.

B. Animals

1. Short-term Exposure

- Groups of 10 male and 10 female F344 rats were exposed to 0, 2,000, 3,500 or 5,000 ppm (0, 4,130, 7,228, or 10,325 mg/m³) chloromethane via inhalation 6 hours/day for 5 days with a 2-day hiatus, followed by 4 more days of treatment (Morgan et al., 1982). Degeneration and necrosis of the proximal convoluted tubule was significantly increased in males at all dose levels and in females at the middle and high-dose levels. Hepatocellular degeneration was significantly increased in females at all dose levels and in males at the middle and high-dose levels. Testicular degeneration was observed at all dose levels. Death, often preceded by ataxia and/or convulsions, occurred in some of the rats exposed to 3,500 or 5,000 ppm (7,228 or 10,325 mg/m³) of chloromethane.

- Groups of five male and five female C3H, C57Bl/6 and B6C3F₁ mice were exposed to 0, 500, 1,000 or 2,000 ppm (0, 1,033, 2,065 or 4,130 mg/m³) of chloromethane via inhalation 6 hours/day for 12 consecutive days (Morgan et al., 1982). Cerebellar degeneration was reported only in strain C57Bl/6 mice administered middle and high doses. Adrenal fatty degeneration was noted in 1,000 and 2,000 ppm groups. Renal degeneration and necrosis were reported in all strains, but only at the highest dose. Hepatocellular degeneration was observed in the C57Bl/6 mice at all dose levels, in B6C3F₁ mice at the highest dose level, and only in male C3H mice at doses of 500 and 2,000 ppm (1,033 and 4,130 mg/m³). All male mice, except C3H, exposed to the highest dose died or became moribund during treatment. Death was often preceded by ataxia. All female mice exposed to the two highest doses, and only male C3H mice exposed to the highest dose, developed hematuria.

- Groups of 12 female C57Bl/6 mice were exposed to 0, 15, 50, 100, 150, 200 or 400 ppm (0, 31, 103, 207, 310, 413 or 826 mg/m³) of chloromethane 22 hours/day for 11 days or to 0, 150, 400, 800, 1,600 or 2,400 ppm (0, 310, 826, 1,652, 3,304 or 4,956 mg/m³) of chloromethane via inhalation 5.5 hours/day for 11 days (Landry et al., 1985). No effects were observed in mice exposed to 50 ppm (103 mg/m³) intermittently. At higher concentrations, histological changes in the cerebellar granule and Purkinje cells, decreased hepatocyte size and focal necrosis in the liver accompanied by increased liver weight and decreased thymus weight were observed. Death was often preceded by ataxia and other neurological disturbances.

- In a study by Smith and von Oettingen (1947a,b), several animal species were exposed to 500, 1,000, 2,000 and 3,000 ppm chloromethane by inhalation for 6 hours/day for up to several months, and were observed for signs of frank toxicity (ataxia and convulsions). Frank effects occurred within a week at 500 ppm in mice, 1,000 ppm in guinea pigs and dogs, and 2,000 ppm in rats, goats and monkeys. The authors concluded that "Young animals responded with slower development of symptoms at the same concentrations where older animals developed symptoms acutely."

2. Dermal/Ocular Effects

• Data regarding dermal exposure to chloromethane could not be located in the available literature.

• Results of ophthalmological examinations revealed no clear cut treatment-related effects in either mice or rats exposed by inhalation to 50 to 1,000 ppm chloromethane for up to 24 months (Pavkov, et al. 1981).

3. Long-term Exposure

• Groups of 10 male and 10 female F344 rats were exposed via inhalation to concentrations of 0, 375, 750 or 1,500 ppm (0, 774, 1,549 or 3,098 mg/m^3) of chloromethane 6 hours/day, 5 days/week for 13 weeks. Food consumption, body weight gain, clinical signs and mortality were measured. Blood and urine samples were analyzed, and ophthalmic examinations were performed. Exposure resulted in decreased body weight gain in all rats exposed to the two highest doses, and increased relative (to body weight) liver weight in female rats exposed to the highest dose (1,500 ppm) (Mitchell et al., 1979).

• Two monkeys exposed via inhalation to 500 ppm (1,033 mg/m^3) of chloromethane for 6 hours/day, 6 days/week, died after 17 weeks of treatment (Smith and von Oettingen, 1947a,b). Death was preceded by progressive debility and terminal unconsciousness.

• Groups composed of 120 male and female F344 rats were exposed via inhalation to 0, 50, 225 or 1,000 ppm (0, 103, 465 or 2,065 mg/m^3) of chloromethane 6 hours/day, 5 days/week ≤ 24 months (Pavkov et al., 1980). The study included ophthalmologic and neurofunctional examinations as well as determination of alterations in survival, body weight, clinical chemistry, hematology, urinalysis, organ weights and histopathology. There was a significant decrease in the growth rate of rats exposed to 1,000 ppm chloromethane. There were also some changes in organ weights at the high-dose level including decreased absolute brain weight in both sexes and decreased absolute and relative (to body weight) testicular weight in males. The testes were the only target organs examined that were considered to have significant chloromethane-induced lesions. At the high dose (1,000 ppm), there was a significant incidence of degeneration and atrophy of the seminiferous tubules and testicular tubules. The NOAEL for rats in this study is 225 ppm.

• Groups composed of 120 male and female B6C3F$_1$ mice were exposed via inhalation to 0, 50, 225 or 1,000 ppm (0, 103, 465 or 2,065 mg/m^3) of chloromethane 6 hours/day, 5 days/week ≥ 24 months (Pavkov et al.,

1981). The study included ophthalmologic and neurofunctional examinations as well as determination of alterations in survival, body weight, clinical chemistry, hematology, urinalysis, organ weights and histopathology. There was treatment-related lethality in female mice exposed to 1,000 ppm. There were also decreased body weights in males and females of the high-dose group. Neurofunctional examination revealed impairment (in clutch response) in both sexes of the high-dose group, which was supported by cerebral lesions noted upon necropsy. Exposure to 1,000 ppm also resulted in increased SGPT levels (male and female) associated with hepatocellular histopathology in males. SGPT levels were increased after 6 and 12 months in mice exposed to 50 and 250 ppm, but this was not associated with any liver lesions. Changes in organ weights in the high-dose group included increased relative heart and liver (female only) weights and decreased absolute brain weights. The only significant change in mice exposed to 225 ppm was an increase in the relative heart weight of females. Histopathological examination revealed several lesions in liver, kidney, cerebellum, spleen and seminiferous tubules in animals exposed to 1,000 ppm. In animals sacrificed at the end of the study (24 months), there was a statistically significant number of renal cortical microcysts in animals exposed to 50 ppm. Thus the LOAEL for this study is 50 ppm based on the incidence of compound-related cortical microcysts observed.

- Groups of 10 male and 10 female F344 rats were exposed via inhalation to 0, 375, 750 or 1,500 ppm (0, 774, 1,549 or 3,098 mg/m³) of chloromethane 6 hours/day, 5 days/week for 13 weeks. Exposure resulted in increased serum glutamic pyruvic transaminase (SGPT) levels and cytoplasmic vacuolization of hepatocytes in male mice exposed to 1,500 ppm (3,098 mg/m³) of chloromethane. Both males and females in the high-dose group had increased liver weights (Mitchell et al., 1979).

4. Reproductive Effects

- Hamm et al. (1985) reported a decrease in male fertility in F344 rats exposed to chloromethane. Rats (40 males and 80 females/group) were exposed via inhalation to 0, 150, 475 or 1,500 ppm (0, 309, 980 or 3,090 mg/m³) chloromethane for 6 hours/day, 5 days/week for 10 weeks prior to mating and 6 hours/day, 7 days/week during 2 weeks of mating. There was a significant ($p < 0.05$) decrease in the number of fertile matings in the males exposed to 475 ppm. Male rats exposed to 1,500 ppm were sterile. There was no effect on female reproductive ability. The decreased male fertility was associated with chloromethane-induced testicular degeneration. After exposure was discontinued, rats exposed to 475 ppm had a full recovery in fertility in 9 weeks, while rats exposed to 1,500 ppm only partially recovered fertility even after an additional 18 weeks of recovery. The NOAEL for reproductive toxicity in this study is 150 ppm. In addition, there was a significant decrease in the M:F ratio and a significant decrease in both male and female growth rates on days 14 to 21 after birth in the F_2 pups exposed to 475 ppm (981 mg/m³) but not 150 ppm (310 mg/m³) of chloromethane.

- Reproductive effects were observed in male F344 rats exposed to 3,000 or 3,500 ppm (6,195 or 7,230 mg/m³) of chloromethane via inhalation 6 hours/day for 5 to 9 days, which included histopathological lesions in the testes, decreased serum testosterone levels, sperm abnormalities and decreased fertility in rats exposed to the two highest doses. No adverse effects were observed in rats exposed to 0 or 1,000 ppm (0 or 2,065 mg/m³) chloromethane via inhalation (Chapin et al., 1984; Working et al., 1985).

- Mitchell et al. (1979) reported that testicular degeneration occurred in rats exposed to 1,000 ppm (2,065 mg/m³) chloromethane via inhalation for 6 to 12 months.

5. Developmental Toxicity

- Groups of 25 pregnant F344 rats were exposed via inhalation to 0, 100, 500 or 1,500 ppm (0, 207, 1,033 or 3,100 mg/m³) of chloromethane 6 hours/day on days 7 to 10 of gestation. Exposure to 1,500 ppm (3,100 mg/m³) resulted in maternal toxicity (decreased body weight gain) and fetotoxicity (decreased fetal weight, decreased female crown-rump length and delayed ossification) (Wolkowski-Tyl et al., 1983b).

- In a study in which pregnant C57B2/6 mice were exposed via inhalation to 250, 500 and 750 ppm (517, 1,033 or 1,552 mg/m³) of chloromethane 6 hours/day on days 6 to 18 of gestation, Wolkowski-Tyl et al. (1983a) reported a significant increase in both the number of fetuses and fetuses/litter with heart defects at levels above 250 ppm. Adverse effects on dams and fetuses were not observed at a concentration of 250 ppm (516 mg/m³) chloromethane.

- Exposure of male F344 rats to 3,000 ppm (6,195 mg/m³) chloromethane via inhalation 6 hours/day for 5 days resulted in a significantly increased preimplantation loss in unexposed dams during weeks 1, 2, 3, 4, 6 and 8 after treatment and a significantly increased postimplantation loss during the first week following treatment (Working et al., 1985).

6. Mutagenicity

- Chloromethane was mutagenic to *Salmonella typhimurium* strains TA100, TA1535 and TA677 with and without metabolic activation (Andrews et al., 1976; Simmon et al., 1977; Simmon, 1978, 1981; Fostel et al., 1985). The lowest dose for which statistical significance was determined was 0.8% (8,000 ppm or 16,560 mg/m³) (Andrews et al., 1976).

- Chloromethane at an atmospheric concentration of 1% (10,000 ppm or 20,700 mg/m³) was mutagenic to TK6 human lymphoid cells *in vitro* and caused an increased incidence of sister chromatid exchange and breakage of DNA strands (Fostel et al., 1985).

7. Carcinogenicity

- Administration of 0, 50, 225 or 1,000 ppm (0, 103, 465 or 2,065 mg/m³) chloromethane via inhalation 6 hours/day, 5 days/week for ≤2 years to groups of 30 male and 30 female B6C3F$_1$ mice resulted in a statistically significant increase in renal tumors among males in the high-dose group. Seventeen renal neoplasms were observed in 13 male mice from the 1,000 ppm group: eight renal cortical adenomas, four adenocarcinomas, two papillary cystadenomas, one papillary cystadenocarcinoma and two tubular cystadenomas. Two renal adenomas also occurred in male mice exposed to 225 ppm. Survival was decreased, but not significantly, among males in the high-dose group (Pavkov et al., 1981).

- No evidence of an oncogenic response was observed in the Pavkov et al. (1981) chronic rat study (see discussion under Long-term Exposure above).

IV. Quantification of Toxicological Effects

A. One-day Health Advisory

The studies by Putz-Anderson et al. (1981a,b) have been selected to serve as the basis for a One-day HA for the 10-kg child because they provide human data and are of appropriate duration. Putz-Anderson et al. (1981b) observed no effects on hand-eye coordination, mental alertness or time discrimination in a group of 12 college students exposed to 200 ppm (413 mg/m³) chloromethane for 3.5 hours. In a similar study by these investigators, a slight but marginally significant (0.04%) impairment was observed in hand-eye coordination, mental alertness and time discrimination in a group of 12 university students exposed to 200 ppm of chloromethane for 3 hours (Putz-Anderson et al., 1981a). Because of the low magnitude of the effect observed in the latter study and no effect observed in the former study when exposure was slightly longer (3.5 hours vs. 3.0 hours), the exposure to 200 ppm (413 mg/m³) for 3.5 hours is judged to be a NOAEL.

The One-day HA for the 10-kg child is calculated as follows:

Step 1: Total Exposed Dose (TED)

$$\text{TED} = \frac{(413 \text{ mg/m}^3)(3.5/24 \text{ hours/day})(20 \text{ m}^3/\text{day})(0.5)}{70 \text{ kg}} = 8.6 \text{ mg/kg/day}$$

where:

413 mg/m^3 = NOAEL from human data.
$3.5/24$ = fractional daily exposure duration.
$20 \text{ m}^3/\text{day}$ = daily ventilation volume in m³ for humans.
0.5 = proportion of exposure dose assumed to be retained (Nolan et al., 1985).
70 kg = average body weight of an adult.

Step 2:

$$\text{One-day HA} = \frac{(8.6 \text{ mg/kg/day}) (10 \text{ kg})}{(10) (1 \text{ L/day})} = 8.6 \text{ mg/L (rounded to } 9{,}000 \text{ } \mu g/L)$$

B. Ten-day Health Advisory

No study was judged to be an adequate basis for a Ten-day HA. It is suggested that the Longer-term HA of 400 μg/L be used as a conservative estimate for a 10-kg child.

C. Longer-term Health Advisory

The study by Repko et al. (1976) has been selected as the basis for the Longer-term HA. In this occupational study, neurological signs were observed at chloromethane average air concentration of 34 ppm (70 mg/m³) after 1 to 26 years of exposure. The use of the Repko study to derive the Longer-term HA is supported by several other animal studies, which also report neurological damage by chloromethane exposure (Smith and von Oettingen, 1947a,b; Morgan et al., 1982; Landry et al., 1985). However, humans seem to be more sensitive than animals to chloromethane neurotoxicity.

The Longer-term HA for a 10-kg child is calculated as follows:

Step 1: Total Exposed Dose (TED)

$$\text{TED} = \frac{(70 \text{ mg/m}^3)(10 \text{ m}^3/\text{day})(5/7)(0.5)}{70 \text{ kg}} = 3.6 \text{ mg/kg/day}$$

where:

70 mg/m^3 = assumed LOAEL in humans.

$10 \text{ m}^3/\text{day}$ = daily ventilation volume of humans in m^3 in an 8-hour workday.

$5/7$ = occupational exposures are 5 workdays/week.

0.5 = proportion of exposure dose assumed to be retained (Nolan et al., 1985).

70 kg = average body weight of an adult.

The Longer-term HA for a 10-kg child is calculated as follows:

$$\text{Longer-term HA (child)} = \frac{(3.6 \text{ mg/kg/day}) (10 \text{ kg})}{(100) (1 \text{ L/day})} = 0.36 \text{ mg/L (rounded to 400 } \mu\text{g/L)}$$

The Longer-term HA for a 70-kg adult is calculated as follows:

$$\text{Longer-term HA (adult)} = \frac{(3.6 \text{ mg/kg/day}) (70 \text{ kg})}{(100) (2 \text{ L/day})} = 1.3 \text{ mg/L (rounded to 1,000 } \mu\text{g/L)}$$

D. Lifetime Health Advisory

While the Pavkov et al. (1980) chronic rat and mouse study was considered as the basis for a Lifetime HA, it was judged that it was prudent to base the HA value on the Repko et al. (1976) study in that this is a human study. For pertinent details concerning the Repko et al. (1976) study, please see the discussion under Longer-term HA.

Step 1: Determination of the Reference Dose (RfD)

$$\text{RfD} = \frac{(3.6 \text{ mg/kg/day})}{(1,000)} = 0.0036 \text{ mg/kg/day}$$

Step 2: Determination of the Drinking Water Equivalent Level (DWEL)

$$\text{DWEL} = \frac{(0.0036 \text{ mg/kg/day}) (70 \text{ kg})}{(2 \text{ L/day})} = 0.13 \text{ mg/L (rounded to 100 } \mu\text{g/L)}$$

Step 3: Determination of the Lifetime Health Advisory

$$\text{Lifetime HA} = \frac{(0.13 \text{ mg/L}) (20\%)}{10} = 0.0026 \text{ mg/L (rounded to 3 } \mu\text{g/L)}$$

E. Evaluation of Carcinogenic Potential

- The International Agency for Research in Cancer has not evaluated the carcinogenic potential of chloromethane.

- Applying the criteria described in the U.S. EPA's Guidelines for Carcinogen Risk Assessment (U.S. EPA, 1986a), chloromethane may be classified in Group C: possible human carcinogen, based on limited positive animal data and positive genotoxic data.

V. Other Criteria, Guidance and Standards

- The 8-hour Time-weighted Average-Threshold Limit Value (TWA-TLV) of 50 ppm (105 mg/m^3), with a 15-minute Short-term Exposure Level (STEL) of 100 ppm (205 mg/m^3) has been recommended by the ACGIH (1980, 1985).

- An ADI of 0.54 mg/kg/day (37.8 mg/day) for a 70-kg human, based on the TLV of 50 ppm and an uncertainty factor of 10 was derived by U.S. EPA (1982). The RfD of 0.0036 mg/kg/day derived in this document supersedes the previously calculated ADI.

VI. Analytical Methods

- Analysis of chloromethane is by a purge-and-trap gas chromatographic procedure used for the determination of volatile organohalides in drinking water (U.S. EPA, 1985a). This method calls for the bubbling of an inert gas through the sample and trapping volatile compounds on an adsorbent material. The adsorbent material is heated to drive off the compounds onto a gas chromatographic column. The gas chromatograph is temperature programmed to separate the method analytes, which are then detected by a halogen-specific detector. This method is applicable to the measurement of chloromethane over a concentration range of 0.1 to 1,500 μg/L. Confirmatory analysis is by mass spectrometry (U.S. EPA, 1985b). The detection limit for confirmation by mass spectrometry has not been determined.

VII. Treatment Technologies

- Available data indicate that air stripping and possibly activated carbon adsorption will remove chloromethane from contaminated water.

- A contaminated ground-water recovery system, consisting of packed column aeration followed by polishing filters and granular activated carbon (GAC), was designed and placed in operation in Hillsborough County, Florida (Higgins et al., 1987). Chloromethane was reportedly removed from an influent concentration of 503 μg/L to below its detection limit. No other operating parameters are provided. McIntyre et al. (1986) report the design and performance of the treatment plant in Hillsborough County, Florida. At an average design flow rate of 150 gpm, the packed column aeration was operated at a hydraulic loading rate of 12 gpm/ft^2 and an air-to-water ratio of approximately 200:1. The three GAC units were operated in series at a hydraulic loading rate of 3 gpm/ft^2. Each contained 8,000 lb of GAC. Each GAC column was designed for an empty bed contact time (EBCT) of 15 minutes and a bed depth of 4.5 feet.

- U.S. EPA (1986c) estimated the feasibility of removing chloromethane from water by packed column aeration employing the engineering design procedure and cost model presented at the 1983 National ASCE Conference on Environmental Engineering. Based on chemical and physical properties and assumed operating conditions, a 99% removal efficiency of chloromethane was reported for a column with a diameter of 5.5 feet and packed with 16.9 feet of 1-inch plastic saddles. The air-to-water ratio required to achieve this degree of removal effectiveness is 11.

- Ashworth et al. (1988) report the results of laboratory measurements of Henry's Law constants for 45 chemicals. These measured values were compared against a batch packed column aeration test. In general, the experimentally determined Henry's Law constants agreed with other reported values. For chloromethane, the Henry's Law constant is 136 atm (20°C).

- The efficiency of powdered activated carbon (PAC) in removing volatile organics, including chloromethane, was investigated at DuPont's Chambers Works Wastewater Treatment Plant (Hutton, 1981). Nuchar SA-15 activated carbon was used at a dosage of 114 mg/L for the treatment of an average 37 million gallons per day (MGD) wastewater containing 169 mg/L soluble TOC. PAC was fed upstream of the aeration chamber designed for 8 hours aeration time. The concentration of chloromethane in the untreated wastewater is not specified. The results show 99 + % removal efficiency of chloromethane by this process. The data presented in this study, however, are not conclusive. Chloromethane is a gas at ambient temperature; therefore, the excellent removal efficiency reported in this study might be the result of treatment by aeration as well as adsorption on activated carbon. Aeration is the probable mechanism of removal.

- Wood and DeMarco (1987) reported the results of a 3-year study at the Preston Water Treatment Plant in Florida. This study investigated the effectiveness of four different types of GAC in removing organic substances from raw, lime-softened and finished water. Filtrasorb 400 was tested in beds 2.5, 5, 7.5 and 10 feet deep with EBCT of 6.2, 12.4, 18.6 and 24.8 minutes, respectively. The other three types of GAC were tested in beds 2.5 feet deep with an EBCT of 6.2 minutes each. Chloromethane was detected in the raw water at a concentration of 0.45 μg/L. Chloromethane was not detected in either lime-softened water or finished water. The GAC usage rates were presented in terms of total purgeable halogenated organic compounds breakthrough and not in terms of chloromethane removal efficiency.

• Evaluation of the physical/chemical properties of chloromethane indicates that GAC may be less effective than air stripping in removing chloromethane from drinking water due to its high vapor pressure and solubility.

VIII. References

ACGIH. 1980. American Conference of Governmental Industrial Hygienists. Documentation of the Threshold Limit Values for substances in workroom air, 4th ed. (Includes supplementary documentation 1981, 1982, 1984, and 1985). Cincinnati, OH: ACGIH. pp. 268–269.

ACGIH. 1985. American Conference of Governmental Industrial Hygienists. Threshold Limit Values for chemical substances in work environment. Adopted by ACGIH with intended changes for 1985–1986. Cincinnati, OH. ACGIH. p. 23.

Ahlstrom, R.C., Jr. and J.M. Steele. 1979. Methyl chloride. In: M. Grayson and D. Eckroth, eds. Kirk-Othmer Encyclopedia of Chemical Technology, 3rd ed., Vol. 5. New York, NY: John Wiley and Sons, Inc., pp. 677–685.

Andrews, A.W., E.S. Zawistowski and C.R. Valentine. 1976. A comparison of the mutagenic properties of vinyl chloride and methyl chloride. Mutat. Res. 40:273–276

Ashworth, R.A., G.B. Howe, M.E. Mullins and T.N. Rogers. 1988. Air-water partitioning coefficients of organics in dilute aqueous solutions. Journal of Hazardous Materials. 18:25–37.

AWWA Research Foundation. 1983. Occurrence and Removal of Volatile Organic Chemicals from Drinking Water. Denver, CO: AWWA.

Browning, E. 1965. Toxicity and Metabolism of Industrial Solvents. New York, NY: Elsevier Publishing Company. pp. 230–241.

Burmaster, D.E. 1982. The new pollution-ground water contamination. Environment. 24:6–13, 33–36.

Bus, J.S. 1978. Disposition of ^{14}C-methyl chloride in Fischer 344 rats after inhalation exposure. Pharmacologist. 20(3):214. Abstract.

Callahan, M.A., M.W. Slimak, N.W. Gabel et al. 1979. Water-related environmental fate of 129 priority pollutants, Vol. II. EPA 440/4-79-029B. Washington, DC: U.S. EPA. pp. 38-1 to 38-9.

Chapin, R.E., R.D. White, K.T. Morgan and J.S. Bus. 1984. Studies of lesions induced in the testes and epididymis of F-344 rats by inhaled methyl chloride. Toxicol. Appl. Pharmacol. 76(2):328-343.

CMR. 1983. Chemical Marketing Reporter. Chemical Profile: Methyl Chloride. New York, NY: Schnell Publishing Co., Inc., pp. 54. March 21.

Crutzen, P.J., I.S.A. Isaksen and J.R. McAfee. 1978. The impact of the chlorocarbon industry on the ozone layer. J. Geophys. Res. 83:345-363.

Dodd, D.E., J.S. Bus and C.S. Barrow. 1982. Nonprotein sulfhydryl alterations in F-344 rats following acute methyl chloride inhalation. Toxicol. Appl. Pharmacol. 62(2):228-236.

Fostel, J., P.F. Allen, E. Bermudez, A.D. Kligerman, J.L. Wilmer and T.R. Skopek. 1985. Assessment of the genotoxic effects of methyl chloride in human lymphoblasts. Mutat. Res. 155(1-2):75-81.

Greenberg, M., R. Anderson, J. Keene, A. Kennedy, G.W. Page and S. Schowgurow. 1982. Empirical test of the association between gross contamination of wells with toxic substances and surrounding land use. Environ. Sci. Technol. 16(1):14-19.

Hamm, T.E., Jr., T.H. Raynor, M.C. Phelps et al. 1985. Reproduction in Fischer-344 rats exposed to methyl chloride by inhalation for two generations. Fund. Appl. Toxicol. 5(3):568-577.

Hansch, C. and A.J. Leo. 1985. Medchem Project. Issue No. 26. Claremont, CA: Pomona College.

Higgins, T.E., J. Dunckel, and B.R. Marshall. 1987. Alternatives to land disposal of hazardous waste and resulting environmental impacts. Presented at the 80th APCA Annual Meeting and Exhibition. June 21-26.

Hutton, D.G. 1981. Removal of priority pollutants with a combined powdered activated carbon-activated sludge process. In: Chemistry in Water Reuse, Vol. 2. pp. 403-428.

Kornbrust, D.J. and J.S. Bus. 1982. Metabolism of methyl chloride to formate in rats. Toxicol. Appl. Pharmacol. 65:135-143.

Kornbrust, D.J. and J.S. Bus. 1983. The role of glutathione and cytochrome P-450 in the metabolism of methyl chloride. Toxicol. Appl. Pharmacol. 67(2):246-256.

Kornbrust, D.J. and J.S. Bus. 1984. Glutathione depletion by methyl chloride and association with lipid preoxidation in mice and rats. Toxicol. Appl. Pharmacol. 72(3):388-399.

Kornbrust, D.J., J.S. Bus, G. Doerjer and J.A. Swenberg. 1982. Association of inhaled [14C]-methyl chloride with macromolecules from various rat tissues. Toxicol. Appl. Pharmacol. 65(1):122-134.

Landry, T.D., T.S. Gushow, P.W. Langvardt, J.M. Wall and M.J. McKenna. 1983. Pharmacokinetics and metabolism of inhaled methyl chloride in the rat and dog. Toxicol. Appl. Pharmacol. 68(3):473–486.

Landry, T.D., J.F. Quast, T.S. Gushow and J.L. Mattisson. 1985. Neurotoxicity of methyl chloride in continuously versus intermittently exposed female C57B1/6 mice. Fund. Appl. Toxicol. 5:87–98.

Lyman, W.J., W.F. Reehl and D.H. Rosenblatt. 1982. Handbook of Chemical Property Estimation Methods. New York, NY: McGraw-Hill, pp. 15–25.

Mabey, W. and T. Mill. 1978. Critical review of hydrolysis of organic compounds in water under environmental conditions. J. Phys. Chem. Ref. Data. 7:383–415.

Mackay, D. and W.Y. Shiu. 1981. A critical review of Henry's Law constants for chemicals of environmental interest. J. Phys. Chem. Ref. Data. 10(4):1175–1199.

McIntyre, G.T., N.N. Hatch, S.R. Gelman and T.J. Peschman. 1986. Design and performance of a groundwater treatment system for toxic organics removal. Journal WPCF. 58(1):41–46.

Mitchell, R.I., K. Pavkov, R.M. Everett and D.A. Holzworth. 1979. A 90-day inhalation toxicology study in rats and mice exposed to methyl chloride. Submitted by DuPont de Nemours and Co., Inc. OTS No. 878211740, Microfiche No. 20586. Washington, DC: OTS.

Morgan, K.T., J.A. Swenberg, T.E. Hamm, Jr., R. Woldowski-Tyl and M. Phelps. 1982. Histopathology of acute toxicity response in rats and mice exposed to methyl chloride by inhalation. Fund. Appl. Toxicol. 2:293–299.

NIOSH. 1984. National Institute for Occupational Safety and Health. Carcinogenic risk assessment for occupational exposure to monochloromethanes. NTIS PB 85–111623.

Nolan, R.J., D.L. Rick, T.D. Landry, L.P. McCarty, G.L. Agin and J.H. Saunders. 1985. Pharmacokinetics of inhaled methyl chloride (CH_3Cl) in male volunteers. Fund. Appl. Toxicol. 5(2):361–369.

Otson, R., D.T. Williams and P.D. Bothwell. 1982. Volatile organic compounds in water at thirty Canadian potable water treatment facilities. J. Assoc. Off. Anal. Chem. 65:1370–1374.

Page, G.W. 1981. Comparison of ground water and surface water for patterns and levels of contamination by toxic substances. Environ. Sci. Technol. 15(12):1475–1481.

Pavkov, K.L., R.I. Mitchell, W.D. Kerns and M.M. Connell. 1980. 24-month interim report on a chronic inhalation toxicology study in rats and mice exposed to methyl chloride. Submitted to the Chemical Industry Institute of Toxicology, NC by Battelle Laboratories, Columbus, OH. TSCA 8d. OTS No. 878212060. Microfiche No. 205952.

Pavkov, K.L., R.I. Mitchell and R.L. Persing. 1981. Final report on a chronic inhalation toxicology study in rats and mice exposed to methyl chloride. Submitted to the Chemical Industry Institute of Toxicology, NC by Battelle Laboratories, Columbus, OH. TSCA 8d. OTS No. 878211741. Microfiche No. 205861.

Perry, R.H. and C.H. Chilton. 1973. Chemical Engineers Handbook, 5th Ed. New York, NY: McGraw Hill.

Putz-Anderson, V., J.V. Setzer, J.S. Croxton and F.C. Phipps. 1981a. Methyl chloride and diazepam effects on performance. Scand. J. Work Environ. Health. 7:8-13.

Putz-Anderson, V., J.V. Setzer, J.S. Croxton and F.C. Phipps. 1981b. Effects of alcohol, caffeine and methyl chloride on man. Psychol. Rep. 48(3):715-725.

Redford-Ellis, M. and A.H. Gowenlock. 1971. Studies on the reaction of chloromethane with preparations of liver, brain and kidney. Acta Pharmacol. Toxicol. 30(1-2):49-58.

Repko, J.D. 1981. Neurotoxicity of methyl chloride. Neurobehav. Toxicol. Teratol. 3(4):425-429.

Repko, J.D. and S.M. Lasley. 1979. Behavioral, neurological, and toxic effects of methyl chloride: a review of the literature. CRC Crit. Rev. Toxicol. 6(4):283-302.

Repko, J.D., P.D. Jones, L.S. Garcia, Jr., E.J. Schneider, E. Roseman and R. Corum. 1976. Behavioral and neurological effects of methyl chloride: behavioral and neurological evaluation of workers exposed to industrial solvents: methyl chloride. DHEW (NIOSH) Publ. No. 77-125. p. 66.

Ruth, J.H. 1986. Odor thresholds and irritation levels of several chemical substances: a review. Am. Ind. Assoc. J. 47(3):142-151.

Scharnweber, H.C., G.N. Spears and S.R. Cowles. 1974. Chronic methyl chloride intoxication in six industrial workers. J. Occup. Med. 16:112-113.

Simmon, V.F. 1978. Structural correlations of carcinogenic and mutagenic alkyl halides. Struct. Correl. Carcinog. Mutagen. DHEW Publ. (FDA) (U.S.). ISS FDA 78-1046. pp. 163-171.

Simmon, V.F. 1981. Applications of the Salmonella/microsome assay. In: Short-Term Tests Chemical Carcinogenesis. pp. 120-126.

Simmon, V.F., K. Kauhanen and R.G. Tardiff. 1977. Mutagenic activity of chemicals identified in drinking water. Dev. Toxicol. Environ. Sci. ISS Prog. Genet. Toxicol. 2:249-258.

Singh, H.B., L.J. Salas and R.E. Stiles. 1983. Methylhalides over the Eastern Pacific (40°N-32°S). J. Geophys. Res. 88:3684-3690.

Singh, H.B., L.J. Salas and R.E. Stiles. 1982. Distribution of selected gaseous organic mutagens and suspect carcinogens in ambient air. Environ. Sci. Technol. 16:872-880.

Sittig, M. 1985. Handbook of Toxic and Hazardous Chemicals and Carcinogens, 2nd Ed. Park Ridge, New Jersey: Noyes Publications.

Smith, W.W. and W.F. von Oettingen. 1947a. The acute and chronic toxicity of methyl chloride. IV. Histopathological observations. J. Ind. Hyg. Toxicol. 29:47–52.

Smith, W.W. and W.F. von Oettingen. 1947b. The acute and chronic toxicity of methyl chloride. II. Symptomatology of animals poisoned by methyl chloride. J. Ind. Hyg. Toxicol. 29:123–128.

Sperling, F., F.J. Macri and W.R. von Oettingen. 1950. No title provided. Arch. Ind. Hyg. Occup. Med. 1:215–244. (Cited in U.S. EPA, 1986b.)

Torkelson, T.F. and V.K. Rowe. 1981. Halogenated aliphatic hydrocarbons containing chlorine, bromine and iodine. In: G. Clayton and F.E. Clayton, eds. Patty's Industrial Hygiene and Toxicology, 3rd ed., Vol. 26. New York, NY: John Wiley and Sons, Inc. pp. 3433–3601.

U.S. EPA. 1977. Computer print-out of non-confidential production data from TSCA inventory. Washington, DC: OPTS, CID, U.S. EPA.

U.S. EPA. 1982. Errata: halomethanes. Ambient Water Quality Criterion for the Protection of Human Health. Prepared by the Office of Health and Environmental Assessment, Environmental Criteria and Assessment Office, Cincinnati, OH for the Office of Water Regulations and Standards, Washington, DC.

U.S. EPA. 1985a. U.S. EPA Method 502.1–Volatile halogenated organic compounds in water by purge and trap gas chromatography. Cincinnati, OH: Environmental Monitoring and Support Laboratory. June 1985 (Revised November 1985).

U.S. EPA. 1985b. U.S. EPA Method 524.1–Volatile organic compounds in water by purge and trap gas chromatography/mass spectrometry. Cincinnati, OH: Environmental Monitoring and Support Laboratory. June 1985 (Revised November 1985).

U.S. EPA. 1986a. Guidelines for Carcinogen Risk Assessment. *Fed. Reg.* 51(185):33992–34003.

U.S. EPA. 1986b. Health and Environmental Effects Profile for Methyl Chloride. Prepared by the Office of Health and Environmental Assessment, Environmental Criteria and Assessment Office, Cincinnati, OH for the Office of Solid Waste and Emergency Response, Washington, DC.

U.S. EPA. 1986c. Economic Evaluation of Chloromethane Removal from Water by Packed Column Air Stripping. Prepared by Office of Water for Health Advisory Treatment Summaries.

Verschueren, K. 1983. Handbook of Environmental Data on Organic Chemicals, 2nd ed. New York, NY: Van Nostrand Reinhold Co. p. 839.

Wolkowski-Tyl, R., M. Phelps and J.K. Davis. 1983a. Evaluation of heart malformations in B6C3F$_1$ mouse fetuses induced by *in utero* exposure to methyl chloride. Teratology. 27:197–206.

Wolkowski-Tyl, R., M. Phelps and J.K. Davis. 1983b. Structural teratogenicity evaluation of methyl chloride in rats and mice after inhalation exposure. Teratology. 27(2):181–195.

Wood, P.R. and J. DeMarco 1987. Treatment of groundwater with granular activated carbon. Journal of Environmental Pathology, Toxicology and Oncology. 7(7/8)241–257.

Working, P.K., J.S. Bus and T.E. Hamm, Jr. 1985. Reproductive effects of inhaled methyl chloride in the male Fischer 344 rat. I. Mating performance and dominant lethal assay. Toxicol. Appl. Pharmacol. 7(1):133–157.

Bromochloromethane

I. General Information and Properties

A. CAS No. 74-97-5

B. Structural Formula

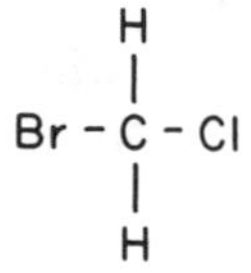

Bromochloromethane

C. Synonyms

- Chlorobromomethane; monochloromonobromomethane; methylene chlorobromide.

D. Uses

- Bromochloromethane is primarily used as a fire-extinguishing fluid, particularly in aircraft and portable extinguishers (Stenger, 1978). It is also used in organic syntheses (Hawley, 1981).

E. Properties (Stenger, 1978; Kudchadker et al., 1979; Hawley, 1981; Wasik et al., 1981; Amoore and Hautala, 1983; Ruth, 1986).

Chemical Formula	CH_2BrCl
Molecular Weight	129.39
Physical State (at 25°C)	Colorless liquid
Boiling Point	67°C
Melting Point	-86.5°C
Density	—
Vapor Pressure (24.05°C)	141.07 mm Hg
Specific Gravity (25°C)	1.93

Water Solubility (25°C)	9,000 mg/L
Log Octanol/Water Partition Coefficient	1.41
Taste Threshold	—
Odor Threshold (water)	34 mg/L
Odor Threshold (air)	2,100 mg/m^3
Conversion Factor (in air)	1 ppm = 5.292 mg/m^3 181 mg/m^3 = 0.1889 ppm

F. Occurrence

- Bromochloromethane was not identified in widespread samples from numerous surveys of chlorinated drinking water, indicating that formation does not occur during chlorination treatment (U.S. EPA, 1985a).

- Bromochloromethane was not identified in ambient air samples in various surveys (U.S. EPA, 1985a).

G. Environmental Fate

- The Henry's Law constant for bromochloromethane was calculated to be 0.0014 atm-m^3/mol, suggesting that volatilization from aqueous environments is both rapid and significant (U.S. EPA, 1985a).

- Rapid biodegradation of bromochloromethane was demonstrated in the Bunch and Chambers static culture flask biodegradability screening test with 5 and 10 mg/L of compound (Tabak et al., 1981). Within 1 week of static incubation at 25°C in the dark, 100% of the compound was degraded.

- Mabey and Mill (1978) estimated a maximum hydrolytic half-life of 44 years for bromochloromethane in water at pH 7 and 25°C, indicating that hydrolysis is not environmentally significant.

- Information pertaining to the photolysis or oxidation of bromochloromethane in water could not be located in the available literature.

- Calculated soil sorption coefficients (K$_{oc}$ values) ranging from 29–137 for bromochloromethane indicate that adsorption on particulates and subsequent precipitation in sediments in aquatic environments are not significant (U.S. EPA, 1985a).

II. Pharmacokinetics

A. Absorption

- Data regarding the oral absorption of bromochloromethane could not be located in the available literature.

- Andersen et al. (1980) reported that the kinetics of absorption of inhaled bromochloromethane was a composite of a slow first order and a saturable uptake process in male Fischer 344 rats and calculated the K_m (the concentration of air at which uptake occurs at one-half the maximum rate) to be 119 ppm (630 mg/m^3) and the V_{max} (the maximum rate of uptake) to be 11.4 mg/kg/hour.

- McDougal et al. (1985) performed an experiment in which male Fischer 344 rats were exposed to 2,500–40,000 ppm (13,235–211,760 mg/m^3) bromochloromethane in air for 4 hours in dermal vapor absorption chambers without significant inhalation exposure. Absorption of bromochloromethane, as measured by blood levels of bromide, was rapid and increased with increasing exposure levels; however, the increase was not linear. A dermal flux of 0.011–0.164 mg/cm^3/ hour was calculated.

B. Distribution

- Data regarding the distribution of bromochloromethane following oral administration could not be located in the available literature.

- In an inhalation study by Svirbely et al. (1947), 2 female dogs, 3 male rabbits and 20 male rats (breed and strains not specified) were separately exposed for 7 hours daily, 5 days/week for 14 weeks to 1,000 ppm (5,294 mg/m^3) bromochloromethane. Both organic and inorganic bromide in the brain and blood of the treated animals were elevated at termination of a 7-hour exposure period, compared with unexposed controls. However, levels of total bromide in the blood varied from 6.0–10.5 times greater than levels in the brain. Levels of organic bromide in the blood varied from 8.0–13.1 times greater than levels in the brain (Svirbely et al., 1947).

C. Metabolism

- When 3.0 mmol bromochloromethane/kg (388.2 mg/kg^3) in corn oil was injected intraperitoneally (i.p.) into male Long-Evans rats weighing 200–350 g, an increase in carboxyhemoglobin levels was observed (Kubic et al., 1974). A maximum saturation of 5% carboxyhemoglobin (absolute levels of carboxyhemoglobin not reported) was measured 4 hours after treatment. This result was interpreted by the investigators as an indication of the production of carbon monoxide during the metabolism of bromochloromethane.

- Svirbely et al. (1947) reported that bromochloromethane is hydrolyzed to inorganic bromide in dogs, rats and rabbits exposed by inhalation to 1,000 ppm (5,294 mg/m^3) 7 hours/day, 5 days/week for 14 weeks. Other metabolites were not reported.

D. Excretion

- Data regarding excretion of bromochloromethane could not be located in the available literature.

III. Health Effects

A. Humans

1. Short-term Exposure

- Rutstein (1963) reported that three men accidentally exposed to bromochloromethane had gastrointestinal disturbances, and central nervous system (CNS) narcosis. The patients recovered with supportive therapy, in particular with the application of artificial respiration during apnea.

2. Long-term Exposure

- Data regarding long-term human exposure to bromochloromethane could not be located in the available literature.

B. Animals

1. Short-term Exposure

- The major response to bromochloromethane exposure is CNS depression (Torkelson and Rowe, 1981).

- Svirbely et al. (1947) reported an oral LD$_{50}$ in Swiss mice of 4,300 mg/kg bw.

- Highman et al. (1948) treated groups of Swiss mice (sex not reported) by stomach tube with single doses of bromochloromethane in olive oil at 0, 500, 3,000 or 4,500 mg/kg. Fatty degeneration of the liver and kidney, as well as focal necrosis and hydropic degeneration of the liver, was observed in the 3,000 and 4,500 mg/kg groups. These changes were most severe after 24 hours and were reversible in mice surviving ≥ 48 hours. No effects were noted at 500 mg/kg.

- Highman et al. (1948) also treated a group of 32 Swiss mice (sex not reported) by gavage with 3,000 mg bromochloromethane/kg/day for 10 consecutive days. Controls were treated with olive oil. Several mice were sacrificed after each of the 10 doses. In mice that died or were killed, fatty degeneration of the liver, kidney and sometimes the heart was observed. These effects were most severe 24–48 hours after the initial dose and became slight after 80 hours. Opacity of the eyes was observed in 5/19 mice surviving ≤ 3 days.

2. Dermal/Ocular Effects

- Dermal application of bromochloromethane (5,000 mg/kg) with occlusion, to the clipped skin of 5 rabbits, resulted in burns and denaturation of the skin with 4/5 rabbits surviving after 24 hours. Application without occlusion had only a defatting effect (Torkelson et al., 1960). A slight decrease in body weight was also observed.

- Application of bromochloromethane to the eyes of rabbits resulted in conjunctival swelling and slight transitory irritation of the cornea (Torkelson et al., 1960).

3. Long-term Exposure

- Torkelson et al. (1960) exposed groups of 40 (20/sex) rats (strain not reported) to 0, 500 or 1,000 ppm (2,646 or 5,292 mg/m^3) and 10 female rats to 370 ppm (1,958 mg/m^3) bromochloromethane by inhalation 7 hours/day, 5 days/week for 4–6 months. In female rats exposed to 370 ppm, body weights, blood urea nitrogen (BUN) levels and gross and microscopic pathology were normal, but liver-to-body weight ratios were increased. In the 500 ppm exposure group, both male and female rats had normal body weights, BUN levels and gross pathology, but liver-to-body weight ratios were increased in rats of both sexes. Histopathologic examination revealed cloudy swelling and vacuolization of hepatocytes.

- Torkelson et al. (1960) exposed groups of 20 (10/sex) guinea pigs (strain not reported) to 0, 500 or 1,000 ppm (2,646 or 5,292 mg/m^3) bromochloromethane by inhalation 7 hours/day, 5 days/week for 4–6 months. At 500 ppm, the body weights of male and female guinea pigs were decreased, and liver-to-body weight ratios were increased; kidney-to-body weight ratios were increased in the males. At 1,000 ppm, both sexes of guinea pigs had decreased body weights and increased liver-and kidney-to-body weight ratios. The females had an increased number of circulating neutrophils.

- Torkelson et al. (1960) also exposed 10 female mice (strain not reported) to 0, 500 or 1,000 ppm (2,646 or 5,292 mg/m^3) bromochloromethane by inhalation 7 hours/day, 5 days/week for 4–6 months. No gross or microscopic pathologic changes were observed at either exposure level. At 1,000 ppm, a slight increase in liver-and kidney-to-body weight ratios was noted.

- Torkelson et al. (1960) exposed four (2/sex) rabbits (strain not reported) to 0, 500 or 1,000 ppm (2,646 or 5,292 mg/m^3) bromochloromethane by inhalation 7 hours/day, 5 days/week for 4–6 months. Two (1/sex) dogs were exposed to 370 ppm (1,958 mg/m^3). Normal body weights and BUN levels were reported for both species.

- MacEwen et al. (1966) exposed 100 (50/sex) albino rats and 8 (4/sex) beagle dogs to 500 or 1,000 ppm (2,646 or 5,292 mg/m^3) bromochloromethane by inhalation 6 hours/day, 5 days/week for 6 months for a total of 124 exposures and observed no adverse effects on clinical chemistry or histopathology. A control group of five male rats and eight (4/sex) dogs was maintained. The only statistically significant effect noted was decreased body weight gain in male rats exposed to 500 and 1,000 ppm as compared with controls.

- Svirbely et al. (1947) exposed 20 male rats, 3 male rabbits and 2 female dogs to 1,000 ppm (5,292 mg/m^3) bromochloromethane 7 hours/day, 5 days/week for 14 weeks. No evidence of toxic response or histopathological changes in liver or kidney was noted, and growth was normal. There was a slight increase of hemosiderin in the spleen and kidney in dogs and the spleen of rats. An accumulation of inorganic bromide in the blood was also noted.

4. Reproductive Effects

- Torkelson et al. (1960) observed decreased (statistical significance not reported) spermatogenesis in 10/10 male guinea pigs and 1/2 male rabbits exposed to 1,000 ppm (5,292 mg/m^3) bromochloromethane by inhalation 7 hours/day, 5 days/week for 4–6 months.

- No studies designed specifically to assess reproductive effects have been reported.

5. Developmental Effects

- Data regarding the developmental effects of bromochloromethane could not be located in the available literature.

6. Mutagenicity

- Bromochloromethane gave a positive dose-related response in the *Salmonella typhimurium* reverse mutation assay in strain TA100 without metabolic activation (Simmon et al., 1977; Simmon, 1978; Simmon and Tardiff, 1978).

- Oserman-Golkar et al. (1983) obtained positive results for bromochloromethane in reverse mutation assays in *S. typhimurium* strains TA100 and TA1535 and in strain W_μ 361089 of *Escherichia coli* and in forward mutation assays in *E. coli* strains Sd-4 and K39. Metabolic activation was not used.

7. Carcinogenicity

- Data on the carcinogenicity of bromochloromethane could not be located in the available literature. Bromochloromethane has not been scheduled for carcinogenicity testing by the National Toxicology Program (NTP, 1987).

IV. Quantification of Toxicological Effects

A. One-day Health Advisory

The study by Highman et al. (1948) has been selected to serve as the basis for the 10-kg child One-day HA because it is a study by a relevant route of exposure (gavage) for an appropriate exposure period (single dose) in a mammalian species (mice) that assessed a parameter of toxicity other than lethality. The liver and kidneys appear to be target organs by both the oral (Highman et al., 1948) and inhalation (Torkelson et al., 1960) routes regardless of duration of exposure. Highman et al. (1948) reported that mice given single gavage doses of 3,000 or 4,500 mg/kg displayed degenerative changes in the liver and kidney. Degenerative liver changes included focal necrosis and hydrophic degeneration. With a single dose of 500 mg/kg, no adverse effects were observed. Therefore, 500 mg/kg/day is a NOAEL for degenerative changes in the liver and kidney in this study.

$$\text{One-day HA} = \frac{(500 \text{ mg/kg/day}) (10 \text{ kg})}{(100) (1 \text{ L/day})} = 50 \text{ mg/L } (50,000 \text{ } \mu\text{g/L})$$

It should be noted that the One-day HA of 50 mg/L exceeds the water odor threshold of 34 mg/L.

B. Ten-day Health Advisory

Since data are insufficient for derivation of a Ten-day HA for bromochloromethane, it is recommended that the Longer-term HA for a 10-kg child of 1 mg/L be used as a conservative approach.

C. Longer-term Health Advisory

No oral studies of sufficient length could be located in the available literature. However, since the toxicological effects are similar in oral and inhalation exposures, it is appropriate to use inhalation studies to derive an oral HA. In an inhalation study, Torkelson et al. (1960) exposed several species to 500 or 1,000 ppm (2,646 or 5,292 mg/m^3) bromochloromethane by inhalation 7 hours/day, 5 days/week for 4–6 months; in addition, a separate group of female rats was exposed to 370 ppm (1,958 mg/m^3) using the same testing regimen. In including female rats, an increased liver-to-body weight ratio was observed at all doses 370 ppm. While these data would suggest that 370 ppm is a LOAEL in female rats, no such observation was made by MacEwen et al. (1966) at 500 ppm in a similar study in rats. However, in that it is not possible to directly compare the Torkelson et al. (1960) study with the MacEwen et al. (1966) study — e.g., species? — it is judged that 370 ppm is a LOAEL in the rat.

Step 1: Total Exposed Dose (TED)

$$\text{TED (rat)} = \frac{(1{,}958 \text{ mg/m}^3)\ (7/24)\ (5/7)\ (0.224 \text{ m}^3/\text{day})\ (0.5)}{0.35 \text{ kg}}$$

$$\text{TED (rat)} = 130.5 \text{ mg/kg/day}$$

where:

$1{,}958 \text{ mg/m}^3$ = LOAEL.
$7/24$ = exposure duration in hours/day.
$5/7$ = exposure is for 5 out of 7 days/week.
$0.224 \text{ m}^3/\text{day}$ = volume inhaled by a rat in 1 day.
0.5 = maximum proportion of exposure dose retained (Anderson et al., 1980).
0.35 kg = weight of a rat.

The Longer-term HA for a 10-kg child is calculated as follows:

$$\text{Longer-term HA} = \frac{(130.5 \text{ mg/kg/day})\ (10 \text{ kg})}{(1{,}000)\ (1 \text{ L/day})} = 1.3 \text{ mg/L (rounded to } 1{,}000\ \mu\text{g/L)}$$

The Longer-term HA for a 70-kg adult is calculated as follows:

$$\text{Longer-term HA} = \frac{(130.5 \text{ mg/kg/day}) (70 \text{ kg})}{(1,000) (2 \text{ L/day})} = 4.57 \text{ mg/L (rounded to}$$
$$5,000 \ \mu\text{g/L)}$$

D. Lifetime Health Advisory

While no chronic studies are available, the Torkelson et al. (1960) — the basis of the Longer-term HA — is an adequate basis for a Lifetime HA.

Step 1: Determination of the Reference Dose (RfD)

$$\text{RfD} = \frac{(130.0 \text{ mg/kg/day})}{(10,000)} = 0.013 \text{ mg/kg/day}$$

Step 2: Determination of the Drinking Water Equivalent Level (DWEL)

$$\text{DWEL} = \frac{(0.013 \text{ mg/kg/day}) (70 \text{ kg})}{(2 \text{ L/day})} = 0.46 \text{ mg/L (rounded to}$$
$$500 \ \mu\text{g/L)}$$

Step 3: Determination of the Lifetime HA

Lifetime HA = $(0.46 \text{ mg/L})(20\%) = 0.092 \text{ mg/L}$ (rounded to 90 μg/L)

E. Evaluation of Carcinogenic Potential

- Data regarding the carcinogenic potential of bromochloromethane could not be located in the available literature. The NTP (1987) has not scheduled this chemical for carcinogenicity testing.

- IARC has not evaluated the carcinogenic potential of bromochloromethane. Applying the criteria described in the U.S. EPA's Guidelines for Carcinogen Risk Assessment (U.S. EPA, 1986b), bromochloromethane may be classified in Group D: Not classifiable. This category is for agents with inadequate animal evidence of carcinogenicity.

V. Other Criteria, Guidance and Standards

- ACGIH (1980, 1985) recommends a TLV of 200 ppm (1,050 mg/m^3) with a STEL of 250 ppm (1,300 mg/m^3). OSHA (1985) promulgated a standard for 200 ppm (1,050 mg/m^3).

VI. Analytical Methods

- Analysis of bromochloromethane is by a purge-and-trap gas chromatographic procedure used for the determination of volatile organohalides in drinking water (U.S. EPA, 1985b). This method calls for the bubbling of an inert gas through the sample and trapping volatile compounds on an adsorbent material. The adsorbent material is heated to drive off the compounds onto a gas chromatographic column. The gas chromatograph is temperature programmed to separate the method analytes, which are then detected by a halogen specific detector. Confirmatory analysis is by mass spectrometry (U.S. EPA, 1985c). The detection limit has not been determined for either method.

VII. Treatment Technologies

- U.S. EPA (1986a) estimated the feasibility of removing bromochloromethane from contaminated ground water by packed column aeration. The estimation employed an engineering design procedure coupled with a cost model to generate cost-optimized design parameters and cost estimates (Cummins and Westrick, 1982). Based on physical and chemical properties and assumed operating conditions, a 50% removal efficiency was reported for a column with a diameter of 7.4 ft, packed with 30 ft of 1-inch plastic saddles and operating at an air-to-water volume ratio of 51:1. Also, an 8.9-ft diameter column packed with 63 ft of 1-inch plastic saddles and operating at an air-to-water volume ratio of 82:1 removed 80% of bromochloromethane. However, it should be emphasized that the equipment sizes and cost estimates were based only on compound properties.

- No data were found for the removal of bromochloromethane from drinking water by activated carbon adsorption. Evaluation of physical/chemical properties indicates that bromochloromethane may not be amenable to removal by activated carbon adsorption due to its high solubility and low molecular weight.

VIII. References

ACGIH. 1980. American Conference of Governmental Industrial Hygienists. Documentation of the Threshold Limit Values for substances in workroom air, 4th ed. Cincinnati, OH: ACGIH. pp. 268–269.

ACGIH. 1985. American Conference of Governmental Industrial Hygienists. Threshold Limit Values for chemical substances in work environment.

Adopted by ACGIH with intended changes for 1985–1986. Cincinnati, OH: ACGIH. p. 2.

Amoore, J.E. and E. Hautala. 1983. Odor as an aid to chemical safety: odor thresholds compared with threshold limit values and volatilities for 214 industrial chemicals in air and water dilution. J. Appl. Toxicol. 3(6):272–290.

Anderson, M.E., M.L. Gargas, R.A. Jones and L.J. Jenkins, Jr. 1980. Determination of the kinetic constants for metabolism of inhaled toxicants *in vivo* using gas uptake measurements. Toxicol. Appl. Pharmacol. 54(1):100–116.

Cummins, M.D. and J.J. Westrick. 1982. Packed column air stripping for removal of volatile compounds. Presented in the Proceedings of the National Conference on Environmental Engineering, ASCE.

Hawley, G.G. 1981. The Condensed Chemical Dictionary, 10th ed. New York, NY: Van Nostrand Reinhold Co. p. 151.

Highman, B., J.L. Svirbely, W.F. von Oettingen, W.D. Alford and L.J. Pecora. 1948. Pathological changes produced by mono-chloro-mono-bromomethane. Am. Med. Assoc. Arch. Pathol. 45:299–305.

Kubic, V.L., M.W. Anders, R.R. Engel, C.H. Barlow and W.S. Caughey. 1974. Metabolisms of dihalomethanes to carbon monoxide I. *In vivo* studies. Drug Metab. Dispos. 2(1):53–57.

Kudchadker, A.P., S.A. Kudchadker, R.P. Shukla and P.R. Patnaik. 1979. Vapor pressures and boiling points of selected halomethanes. J. Phys. Chem. Ref. Data. 8(2):499–517.

Mabey, W. and T. Mill. 1978. Critical review of hydrolysis of organic compounds in water under environmental conditions. J. Phys. Chem. Ref. Data. 7(2):383–415.

MacEwen, J.D., J.M. McNerney, E.H. Vernot and D.T. Harper. 1966. Chronic inhalation toxicity of bromochloromethane. J. Occup. Med. 8:251–256.

McDougal, J.N., G.W. Jepson, H.J. Clewell, III and M.E. Andersen. 1985. Dermal absorption of dihalomethane vapors. Toxicol. Appl. Pharmacol. 79:150–158.

NTP. 1987. National Toxicology Program. Toxicology research and testing program. Management Status Report 10/13/87. Research Triangle Park, NC: NTP.

OSHA. 1985. Occupational Safety and Health Administration. Code of Federal Regulations. 29 CFR 1910.1000.

Osterman-Golkar, S., S. Hussain, S. Walles, B. Anderstam and K. Sigvardsson. 1983. Chemical reactivity and mutagenicity of some dihalomethanes. Chem. Biol. Interact. 46(1):121–130.

Perry, R.H., and C.H.Chilton. 1973. Chemical Engineers Handbook, 5th ed. New York, NY: McGraw Hill Book Company.

Ruth, J.H. 1986. Odor thresholds and irritation levels of several chemical substances: a review. Am. Ind. Hyg. Assoc. J. 47:A142-A151.

Rutstein, H.R. 1963. Acute chlorobromomethane toxicity. Arch. Environ. Health. 7(4):440–444.

Simmon, V.F. 1978. Structural correlations of carcinogenic and mutagenic alkyl halides. DHEW Publ. U.S. FDA, FDA 78–1046, Struct. Correl. Carcinog. Mutagen. pp. 163–171.

Simmon, V.F. and R.G. Tardiff. 1978. The mutagenic activity of halogenated compounds found in chlorinated drinking water. In: R.L. Jolley, H. Gorchev and D.H. Hamilton, Jr., eds. Water Chlorination: Environmental Impact and Health Effects, Vol. 2, Environmental Impact of Water Chlorination: Proceedings of the 2nd Conference, Gatlinburg, TN. October 31- November 4, 1977. Ann Arbor Science Publishers, Inc., Ann Arbor, MI. pp. 417–431.

Simmon, V.F., K. Kauhanen and R.G. Tardiff. 1977. Mutagenic activity of chemicals identified in drinking water. 2nd Int. Conf. Environmental Mutagens, Edinburgh, Scotland, July.

Stenger, V.A. 1978. Bromine Compounds. In: M. Grayson and D. Eckroth, eds. Kirk-Othmer Encyclopedia of Chemical Technology, Vol. 4, 3rd ed. John Wiley and Sons, Inc., NY. pp. 252–269.

Svirbely, J.L., B. Highman, W.C. Alford and W.F. von Oettingen. 1947. The toxicity and narcotic action of mono-chloro-mono-bromomethane with special reference to inorganic and volatile bromide in blood, urine and brain. J. Ind. Hyg. Toxicol. 29:382–389.

Tabak, H.H., S.A. Quave, C.I. Mashni and E.F. Barth. 1981. Biodegradability studies with organic priority pollutant compounds. J. Water Pollut. Control Fed. 53(10):1503–1518.

Torkelson, T.R. and V.K. Rowe. 1981. Halogenated aliphatic hydrocarbons. In: G. Clayton and F.E. Clayton, eds. Patty's Industrial Hygiene and Toxicology, Vol. 2B, 3rd ed. New York, NY: John Wiley and Sons, Inc. pp. 3455–3459.

Torkelson, T.R., F. Oyen and V.K. Rowe. 1960. The toxicity of bromochloromethane (methylene chlorobromide) as determined on laboratory animals. Am. Ind. Hyg. Assoc. J. 21(4):275–286.

U.S. EPA. 1985a. U.S. Environmental Protection Agency. Health and environmental effects profile for bromochloromethanes. Prepared by the Office of Health and Environmental Assessment, Environmental Criteria and Assessment Office, Cincinnati, OH for the Office of Solid Waste and Emergency Response, Washington, DC.

U.S. EPA. 1985b. U.S. Environmental Protection Agency. Method 502.1 —

Volatile halogenated organic compounds in water by purge and trap gas chromatography. Cincinnati, OH: Environmental Monitoring and Support Laboratory. June 1985 (Revised November 1985).

U.S. EPA. 1985c. U.S. Environmental Protection Agency. Method 524.1 – Volatile organic compounds in water by purge and trap gas chromatography/mass spectrometry. Cincinnati, OH: Environmental Monitoring and Support Laboratory. June 1985 (Revised November 1985).

U.S. EPA. 1986a. U.S. Environmental Protection Agency. Economic evaluation of bromochloromethane removal from water by packed column air stripping. Prepared by Office of Water for Health Advisory Treatment Summaries.

U.S. EPA. 1986b. U.S. Environmental Protection Agency. Guidelines for Carcinogen Risk Assessment. *Fed. Reg.* 51(185):33992–34003.

Verschueren, K., 1977. Handbook of Environmental Data on Organic Chemicals. New York, NY: Van Nostrand Reinhold Co.

Wasik, S.P., Y.B. Tewari, M.M. Miller and D.E. Martire. 1981. Octanol/ water partition coefficients and aqueous solubilities of organic compounds. MBSIR81-2406. Washington, DC: U.S. Department of Commerce, National Bureau of Standards.

Weast, R.C. and M. Astle, eds. 1981–82. CRC Handbook of Chemistry and Physics, 62nd ed. Boca Raton, FL: CRC Press Inc.

1,3,5-Trichlorobenzene

I. General Information and Properties

A. CAS No. 108–70–3

B. Structural Formula

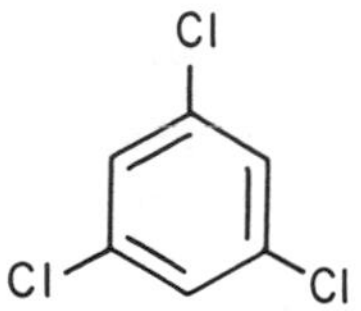

1,3,5-Trichlorobenzene

C. Synonyms

- TCB, sym-trichlorobenzene, s-trichlorobenzene, TCBA.

D. Uses

- 1,3,5-Trichlorobenzene is used as a chemical intermediate; in the synthesis of explosives; in pesticides; and in electrical insulation material (U.S. EPA, 1986a).

E. Properties (U.S. EPA, 1986a; E.I. DuPont deNemours & Co., Inc., 1966)

Chemical Formula	$C_6H_3Cl_3$
Molecular Weight	181.46
Physical State	White crystals/needles
Boiling Point	208.4°C
Melting Point	63.4°C
Density (64°C)	1.39 g/mL
Vapor Pressure (25°C)	0.15 mm Hg
Specific Gravity	—
Water Solubility (25°C)	6.6 mg/L

Log Octanol/Water
   Partition Coefficient   —
Taste Threshold   —
Odor Threshold   6 ppm (approximate)
Conversion Factor   —

F. Occurrence

• Exposure levels of trichlorobenzene isomers in drinking water, food and air are summarized as follows: drinking water, 0.0001–450.0 μg/L; diet, unknown; and air, 0.0081–52.0 μg/m^3. Data are for three trichlorobenzene isomers (1,2,4-; 1,3,5-; 1,2,3-) combined; however, the majority of the data are for 1,2,4-trichlorobenzene. The drinking water and air exposure levels are presented in ranges to reflect the uncertainty arising from the limited amount of monitoring data available. The upper bound for each route should be regarded as a general indicator of exposures, if an individual was exposed to some of the higher reported levels. Air exposure levels are based on the monitoring data presented for urban/ suburban areas, which, in addition to comprising the majority of sampling areas, are more representative of the typical inhalation levels. The U.S. Food and Drug Administration's (FDA's) Total Diet Study data were available for unspecified isomers of trichlorobenzene and, specifically, for 1,2,3-trichlorobenzene. Of the 12 representative food groups identified, only 2 groups had available data represented (one for each previously mentioned isomer group). It was determined, therefore, that since this represents less than 75 percent of the food groups, it is not an adequate representation for total exposure calculations. Consequently, dietary exposure could not be evaluated (U.S. EPA, 1988).

• Available information on the occurrence of trichlorobenzene isomers in drinking water, food and air is insufficient to determine the national distribution of intake by any of the three routes of exposure. While the above data indicate that ambient air exposure can result in higher intakes than food or water, this should be taken only as a relative indication of the potential for intake from ambient air. The number of individuals who actually receive inhalation exposures that are greater than drinking water or dietary exposures is unknown. Therefore, the relative significance of intakes by these routes of exposure for the population as a whole cannot be determined (U.S. EPA, 1988).

G. Environmental Fate

- The persistence of trichlorobenzenes in soil and the only moderate affinity for soil particles pose a potential threat to ground waters underlying sandy soils low in organic carbon content. Terrestrial bioaccumulation is not likely to be significant due to the rapid metabolism of trichlorobenzenes by higher organisms and to the correspondingly more soluble phenols which are readily excreted. Once present in ground waters, advection is expected to be the only significant localized removal process, although anaerobic biodegradation cannot be ruled out. In surface waters, volatilization is the primary route of loss, but it does not occur so rapidly as to preclude contamination of drinking water intakes up to 100 kilometers downstream. Fish bioconcentration factors range up to about 4,000, suggesting that consumption of fish in equilibrium with contaminated water could account for a significant fraction of total exposure under some circumstances (U.S. EPA, 1988).

II. Pharmacokinetics

A. Absorption

- Parke and Williams (1960) administered a single oral dose of 1,3,5-trichlorobenzene at 500 mg/kg as an aqueous suspension to two female Chinchilla rabbits. For the 8 to 9 days after administration, expired air and urine were collected and body contents were examined. Expired air accounted for an average of 11%, urine 9% and the body, including the pelt and depot fat, 30% of the administered dose with a total absorption of approximately 50%. No data on the rate of absorption were available.

B. Distribution

- In the 8 to 9 days following the administration of a single oral dose of 500 mg/kg of 1,3,5-trichlorobenzene as an aqueous suspension to two female Chinchilla rabbits, an average of 20% of the administered dose was detected in gut contents, 5% in the pelt, 5% in depot fat and 21% in the carcass (Parke and Williams, 1960).

- Black et al. (1983) and Ruddick et al. (1983) reported that gas chromatographic (GC) analysis of tissue residues from pregnant rats administered 1,3,5-trichlorobenzene by gavage, on days 6 through 15 of gestation, at levels of 150, 300 or 600 mg/kg bw revealed the presence of the unchanged isomer only in the maternal fat at all dosage levels.

C. Metabolism

- Following the administration of a single oral dose of 500 mg/kg of 1,3,5-trichlorobenzene in arachis oil to female Chinchilla rabbits, Jondorf et al. (1955) reported that 20% of the dose was excreted as glucuronide and 3% as sulfuric acid conjugates; no mercapturic acid was found. Some unchanged 1,3,5-trichlorobenzene was present in the feces while 2,4,6-trichlorophenol was the only phenol detected in the urine.

- Parke and Williams (1960) reported that 1,3,5-trichlorobenzene was not readily metabolized, with the major portion (approximately 50%) of a dose of 500 mg/kg administered as an aqueous suspension to two female Chinchilla rabbits remaining unchanged in the gut contents and tissues 8 to 9 days after dosing. Approximately 4% of the gut contents of one rabbit was identified as a monochlorobenzene. For the first 3 days following dosing, mainly 2,4,6-trichlorophenol was found in the urine along with a minor amount of monochlorophenols. From the fourth through the ninth days, monochlorophenols predominated. 4-Chlorophenol and 2,4,6-trichlorophenol were identified. Approximately 1% of the dose was identified as 4 chlorocatechol. Less than 10% of the dose was excreted as 2,4,6-trichlorophenol.

- Following a single intraperitoneal injection of 60 to 75 mg/kg of 1,3,5-trichlorobenzene in vegetable oil to male rabbits, the major metabolites included 2,3,5-and 2,4,6-trichlorophenol with a third metabolite tentatively identified as a dichlorobenzene with two hydroxyl and one methoxyl substituents (Kohli et al., 1976).

D. Excretion

- Jondorf et al. (1955) reported on the recovery of monophenols from the urine of rabbits given a single oral dose of 500 mg/kg of 1,3,5-trichlorobenzene. A total of 9% of the administered dose was collected over a 5-day recovery period.

- Parke and Williams (1960) reported that 8 to 9 days after a single oral dose of 500 mg/kg of 1,3,5-trichlorobenzene was administered as an aqueous suspension to two female Chinchilla rabbits, an average of 9% of the dose was excreted in the urine, 7% in the feces, and 11% in the expired air, and approximately 50% remained in the body, including gut contents, pelt and depot fat.

III. Health Effects

A. Humans

- No specific data are available on the effects of 1,3,5-trichlorobenzene exposure in humans. It has been reported that exposure to trichlorobenzene (isomers not specified) as a laundry soaking solution resulted in the development of aplastic anemia in a 68-year-old woman, while a 60-year-old man developed anemia after exposure to DDT as well as mono-, di- and trichlorobenzenes for over 30 years (Girard et al., 1969, as cited in U.S. EPA, 1986a).

B. Animals

1. Short-term Exposure

- No data on the oral LD_{50} for 1,3,5-trichlorobenzene in any species were found in the available literature.

- In a study conducted for E.I. DuPont deNemours & Co., Inc. (1966), groups of six mice per dose level were injected intraperitoneally with a single dose of 1,3,5-trichlorobenzene in dimethyl sulfoxide and observed for a period of 2 weeks. No deaths occurred at levels up to 239 mg/kg nor were adverse symptoms observed. The insolubility of the trichlorobenzene prevented the administration of higher doses.

- Six mice were exposed to a saturated vapor of 1,3,5-trichlorobenzene at a concentration of approximately 6 ppm for one 7-hour period. No deaths or gross physiological changes were noted during the 2-week observation period (E.I. DuPont deNemours & Co., Inc., 1966).

- Male Holtzman rats received a single intraperitoneal injection of 5 mM/kg (907 mg/kg) of 1,3,5-trichlorobenzene as a 50% solution in sesame oil 24 hours before cannulation of the femoral vein and common bile duct. A significant increase in bile duct-pancreatic fluid (BDPF) flow was observed. Treatment also resulted in an elevation of serum glutamic pyruvic transaminase (SGPT) activity and a decrease in BDPF protein concentration (Yang et al., 1979).

- Ariyoshi et al. (1975) administered 1,3,5-trichlorobenzene orally to groups of two to six female Wistar rats at a dose of 250 mg/kg/day for 3 days and reported that the microsomal preparations showed increased levels of proteins, phospholipids and cytochrome P-450 content and enhanced the activities of aniline hydroxylase, aminopyrine demethylase and δ-aminolevulinic acid synthetase.

- Carlson (1977) demonstrated the effects of 1,3,5-trichlorobenzene on xenobiotic metabolism. Rats, orally administered 1,3,5-trichlorobenzene at doses of 100 to 200 mg/kg/day for 7 days, demonstrated significant increases in 0-ethyl-p-nitro-phenyl phenylphosphonothionate detoxification, UDP glucuronyltransferase and cytochrome c reductase and a significant decrease in hepatic glucose-6-phosphatase. No significant effects on benzopyrene hydroxylase, azoreductase and serum isocitrate dehydrogenase were evident at 200 mg/kg/day.

- Black et al. (1983) and Ruddick et al. (1983) administered by gavage 0, 150, 300 or 600 mg/kg bw of 1,3,5-trichlorobenzene to pregnant rats on days 6 through 15 of gestation. Thyroid and liver lesions were observed in the adult animals along with significant decreases in erythrocytes, hemoglobin and hematocrit. Doses at which these effects occurred were not specified in abstracts.

- Carlson et al. (1979) and Carlson (1980) demonstrated that oral treatment of male Sprague-Dawley rats treated orally for 14 days with 0.1 mM/kg/day of 1,3,5-trichlorobenzene (approximately 18.2 mg/kg/day in corn oil) had no significant effect on acetanilide hydroxylase, acetanilide esterase and hepatic arylesterase. Pretreatment with 181.5 mg/kg/day of 1,3,5-trichlorobenzene resulted in the induction of procaine esterase.

- Sasmore et al. (1981, 1983) exposed 20 male and 20 female CD (outbred albino) rats per dose level to 1,3,5-trichlorobenzene vapor at nominal doses of 0, 10, 100 or 1,000 mg/m^3 (actual measured dose levels were 0, 9.4, 97 or 961 mg/m^3) for 6 hours/day, 5 days/week for up to 13 weeks. A dried red material was often seen on the faces of the rats in all groups, notably higher in frequency at the high-dose level, and was attributed by the authors to stress. Five rats/sex/dose were sacrificed after 4 weeks of exposure. A small increase in liver-to-body weight ratios in males receiving the high dose was not statistically significant. No effects were seen on the activities of glucose-6-phosphatase and NADPH cytochrome c reductase nor on liver protein content. No significant differences in body weight, food intake, blood chemistry or hematologic parameters were found.

2. Dermal/Ocular Effects

- Data regarding the dermal/ocular effects of 1,3,5-trichlorobenzene could not be located in the available literature.

3. Long-term Exposure

- Sasmore et al. (1981, 1983) exposed 20 male and 20 female CD (outbred albino) rats per dose level to 1,3,5-trichlorobenzene vapor at nominal doses of 0, 10, 100 or 1,000 mg/m^3 (actual measured doses were 0, 9.4, 97 or 961 mg/m^3) for 6 hours/day, 5 days/week for 13 weeks. A dried red material was often seen on the faces of rats in all groups, notably higher in frequency and severity at the high-dose level, and was attributed by the authors to stress. However, this might have been a treatment-related response in the high-dose rats, although its relationship to porphyria (if the red residue is the result of lacrimation, which is rich in porphyrins) is speculative. No significant differences in body weight, food intake, blood chemistry and hematologic parameters were found. Organ weights and ratios did not differ from controls. Urinary porphyrin levels were increased at the 961 mg/m^3 dose level but were not statistically significant, with high standard deviations. The authors acknowledged this result as evidence of an effect but felt that the large standard deviations detract from the importance of it. A squamous metaplasia and focal hyperplasia of the respiratory epithelium of the nasal passages were found in three rats receiving the high dose. Based on indication of effects on urinary porphyrins and respiratory epithelium, a NOAEL of 97 mg/m^3 was concluded.

4. Reproductive Effects

- Data regarding the reproductive effects of 1,3,5-trichlorobenzene could not be located in the available literature.

5. Developmental Effects

- Following the oral administration of 0, 150, 300 or 600 mg/kg bw of 1,3,5-trichlorobenzene to pregnant Wistar rats on days 6 through 15 of gestation, examination of the pups for visceral and skeletal effects indicated that the isomer was not teratogenic. Mild osteogenic alterations were observed but the dose level(s) were not specified (Black et al., 1983 and Ruddick et al., 1983).

6. Mutagenicity

- Data regarding the mutagenic effects of 1,3,5-trichlorobenzene could not be located in the available literature.

7. Carcinogenicity

- Data regarding the carcinogenic effects of 1,3,5-trichlorobenzene could not be located in the available literature.

IV. Quantification of Toxicological Effects

A. One-day Health Advisory

No appropriate data were found in the available literature to calculate a One-day HA. Data from studies of 1 to 3 days duration identify no effects in mice dosed intraperitoneally at doses up to 239 mg/kg or by inhalation at 6 ppm. In contrast, studies in rats dosed orally and intraperitoneally at levels ranging from 37 to 250 mg/kg indicate effects, principally on enzymatic activities, over the total range of doses used. Neither a NOAEL nor a LOAEL can be determined from these available data. Therefore, the Longer-term HA value of 0.6 mg/L for the 10-kg child can serve as a conservative estimate of an exposure which would be considered adequately protective over a 1-day exposure period.

B. Ten-day Health Advisory

No appropriate data were found in the available literature to calculate a Ten-day HA. Data from oral studies of 7 to 14 days duration in rats at doses ranging from 18.2 to 600 mg/kg/day produced significant effects, generally related to enzyme induction, at all levels tested. Neither a NOAEL nor a LOAEL can be determined from these available data. Therefore, the Longer-term HA value of 0.6 mg/L for the 10-kg child can serve as a conservative estimate of an exposure which would be considered adequately protective over a 10-day exposure period.

C. Longer-term Health Advisory

The study by Sasmore et al. (1981, 1983) in which male and female CD rats were exposed to air containing 0, 9.4, 97 or 961 mg/m^3 of 1,3,5-trichlorobenzene for 6 hours/day, 5 days/week for 13 weeks resulted in the development of squamous metaplasia and focal hyperplasia of the respiratory epithelium as well as an indication of porphyria (increases in the urinary porphyrin levels) in rats receiving the high dose. No significant effects on body weight, food consumption, hematological and clinical chemistry parameters or methemoglobin levels were demonstrated at any dose level. A NOAEL of 97 mg/m^3 was concluded from this study. A Longer-term HA based on these data can be calculated as follows:

Determination of Total Absorbed Dose (TAD)

$$TAD = \frac{(97 \text{ mg/m}^3)(1 \text{ m}^3/10^6 \text{ mL})(185 \text{ mL/min})(60 \text{ min/hr})(6 \text{ hrs})(5/7)(0.5)}{(0.4 \text{ kg})}$$

$$= 5.77 \text{ mg/kg/day}$$

where:

97 mg/m^3 = NOAEL for development of respiratory epithelial lesions and effects on urinary porphyrin levels in rats.

$1 \text{ m}^3/10^6 \text{ mL}$ = conversion factor.

185 mL/min = estimated minute volume for a 400 gram rat (Brattlin, 1986).

60 min/hr = conversion factor.

6 hr = exposure duration per day.

0.5 = fraction of test substance shown to be absorbed.

$5/7$ = conversion of 5-day dosing regimen to full 7-day week.

0.4 kg = assumed body weight of an adult rat (Lehman, 1959).

Determination of Longer-term Health Advisory (HA)

For a 10-kg child:

$$\text{Longer-term HA} = \frac{(5.77 \text{ mg/kg/day}) (10 \text{ kg})}{(100) (1 \text{ L/day})} = 0.577 \text{ mg/L (rounded to } 600 \text{ } \mu\text{g/L)}$$

For a 70-kg adult:

$$\text{Longer-term HA} = \frac{(5.77 \text{ mg/kg/day}) (70 \text{ kg})}{(100) (2 \text{ L/day})} = 2.02 \text{ mg/L (rounded to } 2,000 \text{ } \mu\text{g/L)}$$

D. Lifetime Health Advisory

No data are available on the lifetime effects of exposure to 1,3,5-trichlorobenzene in any animal species. However, the study of Sasmore et al. (1981, 1983) which was used to calculate the Longer-term HA may be used for the Lifetime HA by adding an additional uncertainty factor of 10 to account for a study of less than lifetime exposure. The Lifetime HA may be calculated as follows:

Step 1: Determination of the Reference Dose (RfD)

$$RfD = \frac{(5.77 \text{ mg/kg/day})}{(1,000)} = 0.00577 \text{ mg/kg/day}$$

Step 2: Determination of the Drinking Water Equivalent Level (DWEL)

$$DWEL = \frac{(0.00577 \text{ mg/kg/day}) (70 \text{ kg})}{(2 \text{ L/day})} = 0.202 \text{ mg/L } (200 \text{ } \mu\text{g/L})$$

Step 3: Determination of the Lifetime Health Advisory

Lifetime HA = (0.202 mg/L) (20%) = 0.0404 mg/L (40 μg/L)

E. Evaluation of Carcinogenic Potential

- Applying the criteria described in EPA's guidelines for assessment of carcinogenic risk (U.S. EPA, 1986b), 1,3,5-trichlorobenzene may be classified in Group D: not classified. This category is for agents with inadequate animal evidence of carcinogenicity. No studies evaluating the carcinogenic potential of 1,3,5-trichlorobenzene could be located in the available literature.

V. Other Criteria, Guidance and Standards

- There are no United States workplace standards for 1,3,5-trichlorobenzene.

VI. Analytical Methods

- Analysis of 1,3,5-trichlorobenzene is by a purge-and-trap gas chromatographic procedure used for the determination of volatile aromatic and unsaturated organic compounds in water (U.S. EPA, 1985a). This method calls for bubbling an inert gas through the sample and trapping volatile compounds on an adsorbent material. The adsorbent material is heated to drive off compounds onto a gas chromatographic column. The gas chromatograph is temperature-programmed to separate the method analytes which are then detected by the photoionization detector. This method is applicable to the measurement of 1,3,5-trichlorobenzene over a concentration range of 0.3 to 1,500 μg/L. Confirmatory analysis is by mass spectrometry (U.S. EPA, 1985b). The detection limit for confirmation by mass spectrometry has not been determined.

VII. Treatment Technologies

- U.S. EPA (1981) reports 100% removal of 1,3,5-trichlorobenzene from a Cincinnati drinking water source contaminated with several low level organics (μg/L) by granular activated carbon at 8 μg/L influent concentration.

- The calculated value for Henry's Law constant of 66 atm (20°C) for 1,3,5-trichlorobenzene would be amenable to air stripping as a removal technology.

- Because the 1,2,3- and 1,2,4-trichlorobenzene isomers are more common than the 1,3,5-isomer, most of the available data are related to these compounds. Based on the similarities in the structure and the chemical/physical properties of 1,3,5-, 1,2,3- and 1,2,4-trichlorobenzene, the effects of treatment on these compounds will probably be very similar. The following information relates to the 1,2,3- and 1,2,4-trichlorobenzene isomers.

- An adsorption isotherm constant of 157 $mgL^{1/n}/gmg^{1/n}$ and an isotherm slope of 0.31 was reported by Dobbs and Cohen (1980) for 1,2,4-trichlorobenzene from bench-scale tests using granular activated carbon.

- Theim et al. (1987) reported approximately 96% removal of 1,2,4-trichlorobenzene at 100 μg/L influent concentration with 50 mg/L powdered activated carbon. They reported identical removal of 1,2,3-trichlorobenzene under the same conditions.

- U.S. EPA (1986c) estimated the design parameters for a 99% removal efficiency for 1,2,3- and 1,2,4-trichlorobenzene to be equal. A 99% removal efficiency will theoretically be achieved with 28.3 feet of 1-inch plastic saddles and a 27 to 28:1 air to water ratio.

VIII. References

Ariyoshi, T., K. Ideguchi, Y. Ishizuka, K. Iwasaki and M. Arakaki. 1975. Relationship between chemical structure and activity. I. Effects of the number of chlorine atoms in chlorinated benzenes on the components of drug-metabolizing system and the hepatic constituents. Chem. Pharm. Bull. 23(4):817–823. (Cited in U.S. EPA, 1986a.)

Black, W.D., V.E.O. Valli, J.A. Ruddick and D.C. Villeneuve. 1983. The toxicity of three trichlorobenzene isomers in pregnant rats. The Toxicologist. 3(1):30. Abstract.

Brattlin, W.S. 1986. Estimation of breathing rates in animals. Letter to Ken Bailey, U.S. EPA (WH-550D). Life Systems, Inc. November 3.

Carlson, G.P. 1977. Chlorinated benzene induction of hepatic porphyria. Experientia. 22(12):1627–1629. (Cited in U.S. EPA, 1986a.)

Carlson, G.P. 1980. Effects of halogenated benzenes on arylesterase activity *in vivo* and *in vitro*. Res. Commun. Chem. Pathol. Pharmacol. 30(2):361–364.

Carlson, G.P., J.D. Dziezak and K.M. Johnson. 1979. Effect of halogenated benzenes on acetanilide esterase, acetanilide hydroxylase and procaine esterase in rats. Res. Commun. Chem. Pathol. Pharmacol. 25(1): 181–184.

Dobbs, R.A. and J.M. Cohen. 1980. Carbon adsorption isotherms for toxic organics. EPA 600/8–80–023. U.S. EPA Office of Research and Development, HERL, Wastewater Research Division, Cincinnati, OH.

E.I. Dupont deNemours & Co., Inc. 1966. Industrial hygiene evaluation of four compounds used in explosives research. Unpublished report prepared by the Los Alamos Scientific Laboratory of the University of California, Los Alamos, NM. EPA Document No. 878211736.

Girard, R., F. Tolot, P. Martin and J. Bourret. 1969. Serious blood disorders and exposure to chlorine derivatives of benzene (a report of 7 cases). J. Med. Lyon. 50(1164):771–773 (French). (Cited in U.S. EPA, 1986a.)

Jondorf, W.R., D.V. Parke and R.T. Williams. 1955. Studies in detoxication. 66. The metabolism of halogenobenzenes, 1,2,3-, 1,2,4- and 1,3,5-trichlorobenzenes. Biochem. J. 61:512–521. (Cited in U.S. EPA, 1986a.)

Kohli, J., D. Jones and S. Safe. 1976. The metabolism of higher chlorinated benzene isomers. Can. J. Biochem. 54(3):203–208. (Cited in U.S. EPA, 1986a.)

Lehman, A.J. 1959. Appraisal of the safety of chemicals in foods, drugs and cosmetics. Assoc. Food Drug Off. U.S., Q. Bull.

Parke, D.V. and R.T. Williams. 1960. Studies in detoxication. LXXXI. The metabolism of halogenobenzenes: (a) penta- and hexachlorobenzenes, and (b) further observations on 1,3,5-trichlorobenzene. Biochem. J. 74:5–9.

Ruddick, J.A., W.D. Black, D.C. Villeneuve and V.E. Valli. 1983. A teratological evaluation following oral administration of trichloro- and dichlorobenzene isomers to the rat. Teratology. 27(2): 73A-74A. Abstract.

Sasmore, D.P. and D. Palmer. 1981. Ninety-day study of inhaled 1,3,5-trichlorobenzene in rats. Prog. rep. Dept. of Energy, Washington, DC. Rep. No. UCRL-15396. NTIS PC A04/MF A01.

Sasmore, D.P., C. Mitoma, C.A. Tyson and J.S. Johnson. 1983. Subchronic inhalation toxicity of 1,3,5-trichlorobenzene. Drug Chem. Toxicol. 6(3):241–258.

Sittig, M. 1985. Handbook of Toxic and Hazardous Chemicals and Carcinogens, 2nd ed. Park Ridge, NJ: Noyes Publications.

Theim et al. 1987. Adsorption of synthetic organic shock loadings. J. of Environ. Eng. 113(6):1302–1318.

U.S. EPA. 1981. U.S. Environmental Protection Agency. Comparison of grab closed-loop-stripping analysis (CLSA) to other trace organic methods. EPA 600/0–81–060. Cincinnati, OH: U.S. EPA.

U.S. EPA. 1985a. U.S. Environmental Protection Agency. Method 503.1–Volatile aromatic and unsaturated organic compounds in water by purge and trap gas chromatography. Cincinnati, OH: Environmental Monitoring and Support Laboratory. June (Revised November 1985).

U.S. EPA. 1985b. U.S. Environmental Protection Agency. Method 524.1.–Volatile organic compounds in water by purge and trap gas chromatography/mass spectrometry. Cincinnati, OH: Environmental Monitoring and Support Laboratory. (Revised November 1985.)

U.S. EPA. 1986a. U.S. Environmental Protection Agency. Drinking water criteria document for trichlorobenzenes. External review draft. Cincinnati, OH: Environmental Criteria and Assessment Office.

U.S. EPA. 1986b. U.S. Environmental Protection Agency. Guidelines for carcinogen risk assessment. *Fed. Reg.* 51(185):33992–34003. September 24.

U.S. EPA. 1986c. U.S. Environmental Protection Agency. Evaluation of 1,2,4-trichlorobenzene removal from water by packed column air stripping. Prepared by Office of Water for Health Advisory Treatment Summaries.

U.S. EPA. 1988. U.S. Environmental Protection Agency. Occurrence and exposure assessment of trichlorobenzene in public drinking water supplies. Preliminary draft. Washington, DC: Office of Drinking Water.

Williams, R.T. 1959. The metabolism of halogenated aromatic hydrocarbons. In: Detoxication Mechanisms, 2nd ed. New York, NY: John Wiley and Sons. pp. 237–277.

Yang, K.H., R.E. Peterson and J.M. Fujimoto. 1979. Increased bile duct-pancreatic fluid flow in benzene and halogenated benzene-treated rats. Toxicol. Appl. Pharmacol. 47(3):505–514.

1,2,4-Trichlorobenzene

I. General Information and Properties

A. CAS No. 120–82–1

B. Structural Formula

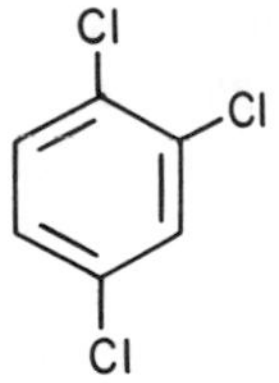

1,2,4-Trichlorobenzene

C. Synonyms

- TCB; asym-trichlorobenzene; Hostetex L-Pec.

D. Uses

- 1,2,4-Trichlorobenzene is used as a solvent in chemical manufacturing; in dyes and intermediates; in dielectric fluid; in synthetic transformer oils; in lubricants; in heat-transfer medium; and in insecticides (U.S. EPA, 1986a).

E. Properties (ACGIH, 1983; Hawley, 1981; U.S. EPA, 1979, 1986a; Verschueren, 1977)

Chemical Formula	$C_6H_3Cl_3$
Molecular Weight	181.46
Physical State	Colorless liquid
Boiling Point	213.5°C
Melting Point	16.95°C
Density (20°C)	1.45 g/mL
Vapor Pressure (25°C)	0.29 mm Hg

Specific Gravity (25°C)   1.4634
Water Solubility (25°C)   34.6 mg/L
Log Octanol/Water
  Partition Coefficient   4.26
Taste Threshold   —
Odor Threshold   3 ppm (approximate)
Conversion Factor   1 ppm = 7.54 mg/m³

F. Occurrence

- Exposure levels of trichlorobenzene isomers in drinking water, food and air are summarized as follows: drinking water, 0.0001 to 450.0 µg/L; diet, unknown; and air, 0.0081 to 52.0 µg/m³. Data are for three trichlorobenzene isomers (1,2,4-; 1,3,5-; 1,2,3-) combined; however, the majority of the data are for 1,2,4-trichlorobenzene. The drinking water and air exposure levels are presented in ranges to reflect the uncertainty arising from the limited amount of monitoring data available. The upper bound for each route should be regarded as a general indicator of exposures, if an individual was exposed to some of the higher reported levels. Air exposure levels are based on the monitoring data presented for urban/suburban areas, which, in addition to comprising the majority of sampling areas, are more representative of the typical inhalation levels. The U.S. Food and Drug Administration's (FDA's) Total Diet Study data were available for unspecified isomers of trichlorobenzene and specifically for 1,2,3-trichlorobenzene. Of the 12 representative food groups identified, only 2 groups had available data represented (one for each previously mentioned isomer group). It was determined, therefore, that since this represents less than 75% of the food groups, it is not an adequate representation for total exposure calculations. Consequently, dietary exposure could not be evaluated (U.S. EPA, 1988).

- Available information on the occurrence of trichlorobenzene isomers in drinking water, food and air is insufficient to determine the national distribution of intake by any of the three routes of exposure. While the above data indicate that ambient air exposure can result in higher intakes than food or water, this should be taken only as a relative indication of the potential for intake from ambient air. The number of individuals who actually receive inhalation exposures that are greater than drinking water or dietary exposures is unknown. Therefore, the relative significance of intakes by these routes of exposure for the population as a whole cannot be determined (U.S. EPA, 1988).

G. Environmental Fate

- The persistence of trichlorobenzenes in soil and the only moderate affinity for soil particles pose a potential threat to ground waters underlying sandy soils low in organic carbon content. Terrestrial bioaccumulation is not likely to be significant due to the rapid metabolism of trichlorobenzenes by higher organisms to the correspondingly more soluble phenols which are readily excreted. Once present in ground waters, advection is expected to be the only significant localized removal process, although anaerobic biodegradation cannot be ruled out. In surface waters, volatilization is the primary route of loss, but it does not occur so rapidly as to preclude contamination of drinking water intakes up to 100 kilometers downstream. Fish bioconcentration factors range up to about 4,000, suggesting that consumption of fish in equilibrium with contaminated water could account for a significant fraction of total exposure under some circumstances (U.S. EPA, 1988).

II. Pharmacokinetics

A. Absorption

- 1,2,4-Trichlorobenzene (TCB) appears to be well absorbed from the gastrointestinal tract of the rat. Following oral administration of 10 mg/kg of ^{14}C-1,2,4-trichlorobenzene (uniformly labeled) in Emulphor EL-620:ethanol:water to Charles River rats (16 males), approximately 84% of the administered dose was recovered in the urine in 24 hours. Recovery in the feces was approximately 11% (Lingg et al., 1982).

- In a similar experiment in two female rhesus monkeys, Lingg et al. (1982) determined that only 38% of an oral dose of 10 mg/kg of ^{14}C-1,2,4-trichlorobenzene (uniformly labeled) in Emulphor EL-620:ethanol:water was recovered in the urine in 24 hours, while 0.1% was recovered in the feces.

- Absorption via the respiratory tract and skin has been demonstrated in toxicity studies (Kociba et al., 1981; Brown et al., 1969) but was not quantitatively determined.

B. Distribution

- Smith and Carlson (1980) studied the distribution of ^{14}C-1,2,4-trichlorobenzene (uniformly labeled) in groups of four male Sprague-Dawley rats on days 1, 6, 11 and 16 after oral dosing with 181.5 mg/kg/day (1 mmol/kg/day; 2 μCi/kg/day) in corn oil for 7 days. Activity on day 1 (dpm/g tissue) was highest in the abdominal fat, followed by the kidneys and liver. By day 6, the levels in the kidney and fat had dropped to approximately 30% of day 1, with the level of activity in the liver being approximately 40%. On day 11, only the abdominal fat and liver showed measurable activities of approximately 20% and 30%, respectively. These activities were maintained to the end of the 16-day study.

C. Metabolism

- Lingg et al. (1982) studied the 24-hour urinary metabolites of a single oral or intravenous dose of 10 mg/kg of ^{14}C-1,2,4-trichlorobenzene in four rats/dosage route. Approximately 60% of the urinary metabolites was accounted for by the 2,3,5- and 2,4,5-isomers of N-acetyl-S-(trichlorophenyl)-L-cysteine. The free 2,4,5- and 2,3,5-isomers of trichlorothiophenol accounted for 33% of the metabolites following oral dosing and 28% following intravenous dosing, while 2,4,5- and 2,3,5-trichlorophenol (TCP) accounted for 1% and 10% of the metabolites, respectively.

- In a study conducted in two rhesus monkeys/dosage route, Lingg et al. (1982) reported that an isomeric pair of 3,4,6-trichloro-3,5-cyclohexadiene-1,2-diol glucuronides accounted for between 48% and 61% of the urinary metabolites following the administration of a single 10 mg/kg dose of ^{14}C-1,2,4-trichlorobenzene via the oral and intravenous routes. Glucuronides of 2,4,5- and 2,3,5-TCP accounted for 14% to 37% of the metabolites while unconjugated TCPs accounted for 1% to 37% of the urinary metabolites.

- Jondorf et al. (1955), as cited in U.S. EPA (1986a), reported that the 5-day urinary metabolites of a single oral dose of 500 mg/kg of 1,2,4-trichlorobenzene administered in arachis oil to three to four Chinchilla rabbits were accounted for by 27% as glucuronide conjugates, 11% as sulfuric acid conjugates and 0.3% as 2,3,4- and 2,4,5-trichlorophenylmercapturic acid. The major phenols were 2,4,5- and 2,3,5-TCP.

- Following a single intraperitoneal injection of 60 to 75 mg/kg of 1,2,4-trichlorobenzene in vegetable oil to male rabbits, the major metabolites were 2,4,5- and 2,3,5-TCP (Kohli et al., 1976).

D. Excretion

- Following the administration of a single oral or intravenous dose of 10 mg/kg of ^{14}C-1,2,4-trichlorobenzene to rats, Lingg et al. (1982) reported that 84% of the oral dose and 78% of the intravenous dose was excreted in the urine by 24 hours while 11% and 7% of the oral and intravenous doses, respectively, were found in the feces.

- When 10 mg/kg of ^{14}C-1,2,4-trichlorobenzene was administered to rhesus monkeys via the oral or intravenous routes, 38% of the oral dose and 22% of the intravenous dose were excreted in the urine after 24 hours with less than 1% appearing in the feces (Lingg et al., 1982).

- During the oral administration of 181.5 mg/kg (1 mmol/kg/day; 2 μCi/kg/day) of ^{14}C-1,2,4-trichlorobenzene in corn oil to Sprague-Dawley rats (4 males) for 7 days, and continuing through 8 days after dosing, Smith and Carlson (1980) reported that fecal elimination accounted for approximately 4% of the total excretion while urinary excretion (approximately 72% of the dose) was still measurable 15 days after the last dose.

- Jondorf et al. (1955) reported the recovery of 42% of a single 500 mg/kg oral dose of 1,2,4-trichlorobenzene in rabbit urine, as trichlorophenols, during a 5-day collection period.

III. Health Effects

A. Humans

- Rowe (1975) reported that an individual exposed to 1,2,4-trichlorobenzene at a level of 3 to 5 ppm developed eye and respiratory irritation.

- Girard et al. (1969) reported that a 68-year-old female exposed to undetermined levels of trichlorobenzene [isomer(s) not specified] as a soaking solution for clothing, developed aplastic anemia. Similarly, a 60-year-old male exposed to DDT as well as mono-, di- and trichlorobenzenes for over 30 years, developed anemia.

B. Animals

1. Short-term Exposure

- Brown et al. (1969) reported an oral LD_{50} for 1,2,4-trichlorobenzene of 756 mg/kg in CFE rats and of 766 mg/kg in CF mice. Death occurred within 3 and 5 days in mice and rats, respectively.

- Male Holtzman rats received a single intraperitoneal (i.p.) injection of 5 mmol/kg of 1,2,4-trichlorobenzene as a 50% solution in sesame oil 24 hours before cannulation of the femoral vein and common bile duct. A significant increase in bile duct-pancreatic fluid (BDPF) flow was observed. Treatment also resulted in a significant increase in bile flow and a decrease in BDPF protein concentration (Yang et al., 1979).

- Robinson et al. (1981) administered i.p. injections of 1,2,4-trichlorobenzene at doses of 0, 250 or 500 mg/kg/day in corn oil to groups of preweanling Charles River female rats (9 to 10/group) at 22, 23 and 24 days of age, and sacrificed the animals at 25 days of age. Significant findings included a decrease in weight of the uterus and an increased liver weight at both dose levels. At the high-dose level, the body weight was significantly decreased and adrenal weight was significantly increased.

- Ariyoshi et al. (1975a) administered 1,2,4-trichlorobenzene orally to groups of two to six female Wistar rats at a dose of 250 mg/kg/day for 3 days and reported that the microsomal preparation showed increased levels of protein, phospholipids and cytochrome P-450 content and enhanced the activity of aniline hydroxylase, aminopyrine demethylase and δ-aminolevulinic acid synthetase.

- Ariyoshi et al. (1975b) also determined the dose response of these effects by administering single oral doses of the 1,2,4-isomer at 0, 125, 250, 500, 750, 1,000 or 1,500 mg/kg. Results, 24 hours after administration, indicated increases in microsomal protein (≥ 750 mg/kg) and cytochrome P-450 content (≥ 250 mg/kg) and the activities of aminopyrine demethylase, aniline hydroxylase and δ-aminolevulinic acid synthetase (≥ 250 mg/kg); while glycogen content decreased (≥ 500 mg/kg).

- Smith and Carlson (1980) administered 181.5 mg/kg/day (1 mmol/kg/day) orally to Sprague-Dawley rats for 7 days in corn oil and measured enzyme recovery at 1, 6, 11 and 16 days after termination of dosing. p-Nitroanisole demethylase activity was significantly elevated through 16 days, O-ethyl-p-nitrophenyl phenylphosphonothionate (EPN) detoxification and cytochrome P-450 through 11 days, and NADPH cytochrome c reductase through 6 days. No increase in induction due to starvation for 4 days was observed, nor was there a significant change in levels of ^{14}C in fat and liver after starvation.

- In a series of experiments (Carlson 1977a, 1980; Carlson and Tardiff, 1976; Carlson et al., 1979), male albino rats were administered 1,2,4-trichlorobenzene in corn oil at doses ranging from 10 to 600 mg/kg/day over a 14-day period. While several microsomal enzymes, including cytochrome c reductase, were induced to varying degrees at dose levels from 10 to 40 mg/kg/day, cytochrome P-450 content was increased (≥ 20 mg/kg/day), and glucose-6-phosphatase (≥ 300 mg/kg/day) and hexabarbital sleeping time (≥ 600 mg/kg/day) were decreased. No effect was seen on activities of serum isocitrate dehydrogenase and arylesterase, nor on hepatic acetanilide hydroxylase. Although liver/body weight ratios were increased at doses of 10, 20 and 40 mg/kg/day, this response was repeated only at the 40 mg/kg/day dose in the 90-day study by Carlson and Tardiff (1976).

- Rimington and Ziegler (1963) determined that oral administration of 1,2,4-trichlorobenzene to three male albino rats at 730 mg/kg/day for 15 days induced porphyria as evidenced by peak elevations in urinary coproporphyrin, uroporphyrin, porphobilinogen and δ-aminolevulinic acid. At doses of 500 mg/kg/day for 10 days (5 rats), peak liver levels of coproporphyrin, protoporphyrin, uroporphyrin and catalase were reached. Glutathione had a protective effect on trichlorobenzene-induced porphyria.

- Kociba et al. (1981) exposed 20 male Sprague-Dawley rats to 1,2,4-trichlorobenzene vapor at 0, 30 or 100 ppm (approximately 0, 223 or 742 mg/m³) for 7 hours/day, 5 days/week for 44 days (30 total exposures). There were no significant effects on body weight, hematology or blood chemistry and gross and microscopic indices. At the high dose, liver and kidney weights were increased, and increased excretion of porphyrin was noted at both dose levels. Exposure of rabbits (four males) and dogs (two males) to the same dose levels revealed only increased liver weights in the dogs.

2. Dermal/Ocular Effects

- The acute percutaneous LD_{50} was determined in CFE rats (4/sex) to be 6,139 mg/kg (Brown et al., 1969) after topical administration of 1,2,4-trichlorobenzene to the shaved dorsolumbar skin, which was then covered with an impermeable dressing. All deaths occurred within 5 days.

- Brown et al. (1969) reported some focal necrosis of the liver in guinea pigs that died after treatment 5 days/week for 3 weeks with 0.5 mL of 1,2,4-trichlorobenzene to the shaved middorsal skin. Some fissuring of the skin was noted in the guinea pigs but little irritation was seen.

- Technical grade 1,2,4-trichlorobenzene was applied at concentrations of 5%, 25% or 100% (0.2 mL/exposure) in petroleum ether to the inner surface of one ear of New Zealand White rabbits (12/level) three times weekly for 13 weeks to assess its chloracnegenic potential. No signs of systemic toxicity related to the exposure were noted. Both the solvent control and the 5% test level developed slight redness, scaling and desquamation by 39 to 40 exposures with no increase in severity thereafter. Moderate to severe irritation characterized by slight to severe erythema, severe scaling, desquamation and encrustation was evident in the 25% and 100% groups along with slight enlargement of the follicles, some hair loss and scarring. There was no evidence of thickening of the ear, exudate or acneform dermatitis. Skin biopsies revealed dermal irritation with slight to moderate acanthosis and hyperkeratosis more pronounced at 25% and 100%. At the 100% level, slight secondary involvement of the follicular sheath was evidenced at necropsy in two rabbits. No other histopathological findings were evident in the visceral organs (Powers et al., 1975).

- Dow Chemical Co. (1958) indicated that eye contact with undiluted 1,2,4-trichlorobenzene or a 10% solution in propylene glycol in rabbits resulted in moderate and immediate pain accompanied by slight conjunctivitis which subsided in 24 hours.

3. Long-term Exposure

- Carlson (1977b) exposed groups of five female rats to oral doses of 1,2,4-trichlorobenzene at 0, 50, 100 or 200 mg/kg/day for 30, 60, 90 or 120 days. After 30 days treatment, a significant increase in liver porphyrins was noted at ≥ 100 mg/kg/day and in urinary porphyrins at 200 mg/kg/day, along with a slight but significant increase in liver weights at the high dose. After 60 days, only the liver weights were increased (at all doses). When 1,2,4-trichlorobenzene was administered for 90 days, a slight but significant increase in liver weight occurred at all dose levels, while liver porphyrins were increased at ≥ 100 mg/kg/day and urinary porphyrins at 200 mg/kg/day. After 120 days exposure, liver porphyrins were significantly increased at ≥ 100 mg/kg/day and urinary porphyrins were significantly increased at 100 mg/kg/day.

- Smith et al. (1978) administered 1,2,4-trichlorobenzene orally to rhesus monkeys (4/group) at doses of 1, 5, 25, 90, 125 or 173.6 mg/kg/day for 90 days. At the high-dose level, two deaths occurred within 20 to 30 days and doses ≥ 90 mg/kg/day were toxic. The high-dosed animals exhibited severe weight loss and predeath tremors. These monkeys also demonstrated elevated blood urea nitrogen (BUN), Na^+, K^+, creatine phosphokinase (CPK), serum transaminases (SGOT, SGPT), lactic dehydrogenase (LDH) and alkaline phosphatase, along with hypercalcemia, hyperphosphatemia and induction of cytochrome P-450. Fatty infiltration was noted in the livers of the high-dosed groups. The NOAEL for this study was 25 mg/kg/day. However, there were weight losses in control animals, and the monkeys were treated with drugs.

- Carlson and Tardiff (1976) orally administered 1,2,4-trichlorobenzene in corn oil to male CD rats (6/group) at levels of 0, 10, 20 or 40 mg/kg/day for 90 days. No consistent effects on weight gain, hemoglobin content and packed cell volume were seen, but at the high-dose level (40 mg/kg/day), a statistically significant increase in liver-to-body weight ratio occurred and persisted throughout a 30-day recovery period. Increases in the following indicators of xenobiotic metabolism occurred after 90 days at the doses indicated: cytochrome c reductase and azoreductase activities (≥ 10 mg/kg/day), cytochrome P-450 levels and EPN detoxification (≥ 20 mg/kg/day) and benzopyrene hydroxylase (two-fold increase, 40 mg/kg/day). The activity of glucuronyltransferase decreased (≥ 10 mg/kg/day). Recovery to normal did not occur after 30 days posttreatment in the levels of cytochrome c reductase and cytochrome P-450 activity. Liver pathology was unremarkable. The NOAEL is 20 mg/kg/day.

- Watanabe et al. (1978) exposed male and female Sprague-Dawley rats to 1,2,4-trichlorobenzene vapor at 0, 3 or 10 ppm (approximately 0, 22.3 or 74.2 mg/m³) for 6 hours/day, 5 days/week for 3 months. Urinary excretion of uroporphyrins was slightly increased at the high dose but returned to control levels in 2 to 4 months post-exposure. The NOAEL for this study is 3 ppm or 22.3 mg/m³.

- 1,2,4-Trichlorobenzene was orally administered to groups of four rhesus monkeys at doses of 1 to 25 mg/kg/day for 120 days. At these doses, there was no significant induction of cytochrome P-450 or P-448 nor was there any effect on weight, hematological or clinical chemistry parameters or physical signs. When a dose of 125 mg/kg/day was administered, a temporary weight loss occurred and one of the four monkeys died; cytochrome P-450 was induced at this dose (Cragg et al., 1978). A NOAEL of 25 mg/kg is indicated.

- In a study in which rats (30/group), rabbits (16/group) and monkeys (8/group) were exposed to 1,2,4-trichlorobenzene vapor at nominal doses of 0, 25, 50 or 100 ppm (equivalent to approximately 1, 190, 375 or 755 mg/m³) for 7 hours/day, 5 days/week for up to 26 weeks, Coate et al. (1977) reported no dose-related effects on general health, weight, chemical or hematological parameters in any species. Rabbits and monkeys showed no adverse effects on the eye and monkeys showed no effects on pulmonary function or operant behavior. Upon necropsy, mild effects due to exposure included hepatocytomegaly (dose-related) and vacuolization of hepatocytes, granuloma formation and biliary hyperplasia of the liver and hyaline degeneration of the kidney at 4 to 13 weeks in rats at most dose levels. These effects were no longer evident in rats treated for 26 weeks. Rabbits and monkeys showed no histopathological effects due to exposure when sacrificed at 26 weeks (no interim necropsy). Based upon the effects on the liver of rats at all dose levels, a NOAEL cannot be established for this study.

4. Reproductive Effects

- Robinson et al. (1981) conducted a multigeneration reproduction study in timed-pregnant Charles River rats (20/group). Upon birth of the F_0 generation and continuing through weaning of the F_2 generation, the rats were administered 1,2,4-trichlorobenzene at 0, 25, 100 or 400 ppm (calculated to average 0, 8.4, 27.8 or 133.4 mg/kg/day at 29 days of age for males and females, and 0, 3.7, 14.8 or 53.6 mg/kg/day for males and 0, 2.5, 8.9 or 33.0 mg/kg/day for females at 83 days of age) in drinking water containing 0.125% Tween 20. Both a water and a 0.125% Tween 20 in drinking water control group were included. Production of the subsequent generation began at approximately day 90 in the F_0 and F_1 generations. Treatment with 1,2,4-trichlorobenzene had no effects on gestation index, fertility, neonate weight, maternal weight, litter size, viability during preweaning or postweaning growth in any generation, nor were effects due to treatment seen on locomotor activity or time of vaginal opening. Significant effects at the high-dose level included an increase in dietary intake in F_0 males at 29 days, and a decrease in water intake in F_0 females at 35 days and in both sexes at 83 days. A significant increase in adrenal weights was seen in both sexes at 95 days of age in the high-dose group of both the F_0 and F_1 animals. A significant increase was also seen in the kidney weight of the Tween 20 control group of the F_1 animals. No treatment-related effects were noted in blood chemistry parameters or gross pathology, or upon histopathological examination of the liver and kidneys in any of the treated animals. A NOAEL of approximately 14.8 mg/kg/day is indicated.

5. Developmental Effects

- Timed-pregnant rats administered 1,2,4-trichlorobenzene orally on days 6 through 15 of gestation at levels of 0, 75, 150 or 300 mg/kg bw indicated maternal toxicity as evidenced by thyroid and liver lesions and significant reductions in erythrocytes, hemoglobin and hematocrit. No teratogenic effects were produced. Mild osteogenic effects were seen in the exposed pups (Black et al., 1983 and Ruddick et al., 1983).

- Maternal toxicity was investigated in a study by Kitchin and Ebron (1983) in timed-pregnant Sprague-Dawley rats (6 or more/group) orally administered 1,2,4-trichlorobenzene in corn oil at doses of 0, 36, 120, 360 or 1,200 mg/kg/day on days 9 through 13 of gestation and sacrificed on day 14. All high-dosed animals were dead by day 3, while the next highest dose level had a mortality rate of 22%. No effects were seen on absolute and relative liver weights and hepatic microsomal protein content. NADPH-cytochrome c reductase activity was increased at the 360 mg/kg/day level. At ≥ 120 mg/kg/day, body weight gain was markedly decreased while the activity of several hepatic microsomal enzymes was significantly increased, including the activity of cytochrome P-450, aminopyrine N-demethylase, ethoxyresorufin O-deethylase, UDP-glucuronyl transferase toward p-nitrophenol and glutathione S-transferase. Hepatic reduced glutathione levels were significantly decreased at the 360 mg/kg/day level. No liver histology was evident at 36 mg/kg/day while 11% (1/9) of the next highest dose showed slight hepatocellular hypertrophy. At 360 mg/kg/day, the hypertrophy was moderate while moderate to moderately severe multifocal necrosis was seen in the rats that died at the highest test dose. Administration of 1,2,4-trichlorobenzene at 360 mg/kg/day resulted in an adverse effect on embryonic development as evidenced by a decrease in head length, crown-rump length, somite number and protein content. Resorptions and abnormalities were not significantly increased. Embryolethality was confined to 3 of the 12 litters examined. This study identified a maternal NOAEL of 36 mg/kg/day while embryonic effects were evident at the only dose examined (360 mg/kg/day).

6. Mutagenicity

- The *Salmonella*/microsomal assay using strains TA1535, 1537, 98 or 100 was negative for mutagenicity when 1,2,4-trichlorobenzene was tested, with and without metabolic activation, at doses ranging from 102 to 1.4×10^{-5} μg/plate. The toxic dose was 1,599 μg/plate. When S-9 extracts were prepared from rats pretreated with 1,2,4-trichlorobenzene, the mutagenicity of 2 aminoanthracene in strains TA1538 and 100 was generally increased (Schoeny et al., 1979).

- No mutagenic activity was seen in the *Salmonella* plate incorporation mutagenesis assay in strains TA98, 100, 1535, 1537 and 1538, with and without metabolic activation with Arochlor 1252 (dose range not specified) (Lawlor et al., 1979).

7. Carcinogenicity

- No adequate study on the carcinogenic potential of 1,2,4-trichlorobenzene was found in the available literature. A 2-year skin painting study conducted by Yamamoto et al. (1982), as cited in U.S. EPA (1986a), in mice indicated only that no single tumor type was increased significantly over control incidence. However, the high dose

(60% in acetone) males had 9 different tumors compared to 2 in both the low (30% in acetone) and control (acetone) groups. In females, these numbers were 11, 3 and 8 for the high, low and control groups, respectively.

IV. Quantification of Toxicological Effects

A. One-day Health Advisory

No appropriate data were found in the available literature to calculate a One-day HA. Therefore, the Longer-term HA value of 0.1 mg/L for the 10-kg child can serve as a conservative estimate of a One-day HA.

B. Ten-day Health Advisory

No appropriate data were found in the available literature to calculate a Ten-day HA. Therefore, the Longer-term HA value of 0.1 mg/L for the 10-kg child can serve as a conservative estimate of a Ten-day HA.

C. Longer-term Health Advisory

Three studies were evaluated as the possible basis for the derivation of Longer-term HAs for a 10-kg child or a 70-kg adult. Two of the studies were oral studies (Carlson and Tardiff, 1976; Carlson, 1977b) and one was an inhalation study (Watanabe et al., 1978).

In the 90-day study by Carlson and Tardiff (1976), the effects of oral dosing of male CD rats (6 animals/group) with 1,2,4-trichlorobenzene in corn oil at 0, 10, 20 or 40 mg/kg/day on weight gain, liver weight, hemoglobin content, packed cell volume and the indicators of xenobiotic metabolism were evaluated. No effects on weight gain and no consistent alteration in hemoglobin content or packed cell volume were observed. At 40 mg/kg, there was a statistically significant increase ($p < 0.05$) in liver-to-body weight ratios that persisted throughout a 30-day recovery period. Following 90-day administration, cytochrome c reductase activity was increased at ≥ 10 mg/kg, with no recovery after 30 days; cytochrome P-450 levels increased at ≥ 20 mg/kg, followed by no recovery; glucuronyltransferase activity decreased at ≥ 10 mg/kg; EPN detoxication increased at ≥ 20 mg/kg; benzopyrene hydroxylase activity increased twofold at 40 mg/kg; and azoreductase activity increased at ≥ 10 mg/kg.

In the Carlson (1977b) study, groups of five female rats were administered daily oral doses of 0, 50, 100 or 200 mg 1,2,4-trichlorobenzene/kg/day in corn oil for 30, 60, 90 or 120 days. Significant increases were observed in liver porphyrins at ≥ 100 mg/kg after 30 days exposure and in urinary porphyrins at 200 mg/kg after 30 days. For the 30-day study, slight but significant increases were also observed in liver weights at 200 mg/kg. When the compound was administered for 60 days, only the liver weights were increased (at all doses). The administration of 1,2,4-trichlorobenzene for 90 days resulted in slight but significant increases in liver weights at ≥ 50 mg/kg, in liver porphyrins at ≥ 100 mg/kg and in urine porphyrins at 200 mg/kg. A significant increase was observed for liver porphyrins when the compound was given at ≥ 100 mg/kg for 120 days. The excretion of δ-aminolevulinic acid and porphobilinogen in the urine was not increased at any dose given for any duration. From the data, the 50 mg/kg/day dose level can be considered a LOAEL.

Watanabe et al. (1978) exposed male and female Sprague-Dawley rats to 1,2,4-trichlorobenzene at 0, 22.3 mg/m^3 (3 ppm) or 74.2 mg/m^3 (10 ppm) for 6 hours/day, 5 days/week for 3 months. The results, which were reported in an abstract, indicated that urinary excretion of porphyrins was slightly increased in the 74.2 mg/m^3 group during exposure, but returned to control range 2 to 4 months postexposure. Since this appeared to be the most sensitive indicator in rats, and exposure to trichlorobenzene at 22.3 mg/m^3 did not cause increased porphyrin excretion, 22.3 m/m^3 was considered a NOAEL for rats by the authors (Watanabe et al., 1978).

The study by Carlson and Tardiff (1976) used the oral route of exposure, which is preferred for deriving drinking water HAs, but this study was primarily a study of 1,2,4-trichlorobenzene's ability to induce xenobiotic metabolizing enzymes and to alter related parameters such as liver weights. The critical adverse effect induced by 1,2,4-trichlorobenzene of porphyria-related effects was not evaluated in this study and, therefore, makes this study inappropriate for deriving HAs. The Carlson (1977b) study also used the preferred oral route of exposure and did evaluate the test animals for the critical effect of porphyria. However, all the experimental animals in the Carlson (1977b) study that received 1,2,4-trichlorobenzene were observed with significant differences from controls during 120 days of exposure, which indicates that only a LOAEL and no NOAEL can be determined from this study. The Watanabe et al. (1978) study used the less desirable inhalation route of exposure, but both the oral and inhalation routes produce similar systemic effects such as porphyria. The Watanabe et al. (1978) study does provide dose-response data identifying NOAEL and LOAEL levels based on increased porphyrin excretion in rats. Therefore, an evaluation of these three studies determined that the Watanabe et al. (1978) study is the most appropriate study to derive Longer-term HAs and the Lifetime DWEL.

The Longer-term HAs for a 10-kg child and a 70-kg adult are calculated as follows:

Determination of Total Absorbed Dose (TAD):

$$\text{TAD} = \frac{(22.3 \text{ mg/m}^3)(6/24)(5/7)(0.279 \text{ m}^3/\text{day})(0.5)}{0.423 \text{ kg}} = 1.31 \text{ mg/kg/day}$$

where:

22.3 mg/m^3 = NOAEL based on the absence of adverse effects in rats (Watanabe et al., 1978).

$6/24$ = exposure duration in hours/day.

$5/7$ = exposure frequency in days/week.

$0.279 \text{ m}^3/\text{day}$ = assumed daily ventilation volume $[(0.5328 \times \text{bw}^{3/4} \text{ (in kg)}) = \text{m}^3/\text{day}.]$

0.423 kg = average body weight of rats at the end of the study.

0.5 = assumed proportion of exposure dose retained.

Determination of Longer-term Health Advisory (HA):

The Longer-term HA for a 10-kg child is calculated as follows:

$$\text{Longer-term HA} = \frac{(1.31 \text{ mg/kg/day}) (10 \text{ kg})}{(100) (1 \text{ L/day})} = 0.131 \text{ mg/L (rounded to } 100 \text{ } \mu\text{g/L)}$$

The Longer-term HA for a 70-kg adult is calculated as follows:

$$\text{Longer-term HA} = \frac{(1.31 \text{ mg/kg/day}) (70 \text{ kg})}{(100) (2 \text{ L/day})} = 0.459 \text{ mg/L (rounded to } 500 \text{ } \mu\text{g/L)}$$

D. Lifetime Health Advisory

As discussed in the Longer-term HA section, the Watanabe et al. (1978) study is the most appropriate to derive a Lifetime DWEL. The Lifetime DWEL for a 70-kg adult is calculated as follows:

Step 1: Determination of the Reference Dose (RfD)

$$\text{RfD} = \frac{(1.31 \text{ mg/kg/day})}{(1,000)} = 0.00131 \text{ mg/kg/day}$$

Step 2: Determination of the Drinking Water Equivalent Level (DWEL)

$$\text{DWEL} = \frac{(1.31 \text{ } \mu\text{g/kg/day}) (70 \text{ kg})}{(2 \text{ L/day})} = 0.0459 \text{ mg/L } (45.9 \text{ } \mu\text{g/L})$$

Step 3: Determination of the Lifetime Health Advisory

Lifetime HA = (0.0459 mg/L) (20%) = 0.009 mg/L (9 μg/L)

E. Evaluation of Carcinogenic Potential

* Applying the criteria described in EPA guidelines for assessment of carcinogenic risk (U.S. EPA, 1986b), 1,2,4-trichlorobenzene may be classified in Group D: not classifiable. This category is for agents with inadequate animal evidence of carcinogenicity. The study of Yamamoto et al. (1982) was not a definitive indicator of carcinogenicity, although the authors indicated an increase in the tumor types in the high-dosed animals (60% in acetone).

V. Other Criteria, Guidance and Standards

* ACGIH (1983) has proposed a ceiling limit of 5 ppm (40 mg/m³) for 1,2,4-trichlorobenzene.

VI. Analytical Methods

* Analysis of 1,2,4-trichlorobenzene is by a solvent extraction gas chromatographic procedure used for the determination of chlorinated hydrocarbons in water samples (U.S. EPA, 1984a). This method requires the extraction of a 1-liter sample with methylene chloride using a separatory funnel. The methylene chloride extract is dried and exchanged to hexane during concentration to a volume of 10 mL or less. A portion of the concentrated sample is injected into a gas chromatograph with an electron capture detector. The reported method detection limit for 1,2,4-trichlorobenzene is 0.05 μg/L. Confirmatory analysis for this contaminant is by gas chromatography/mass spectrometry (U.S. EPA, 1984b). The detection limit for confirmation by mass spectrometry is 1.9 μg/L.

VII. Treatment Technologies

* Available data indicate that granular activated carbon (GAC) and powdered activated carbon (PAC) adsorption, air stripping, reverse osmosis (RO) and possibly ozonation will remove 1,2,4-trichlorobenzene from contaminated water.

* Love and Miltner (1985) report on field and laboratory studies undertaken to determine the effectiveness of GAC in removing 1,2,4-trichlorobenzene. Filtrasorb-300 exhibited an adsorptive capacity of 128

mg 1,2,4-trichlorobenzene/gm carbon at an equilibrium concentration of 500 μg/L.

- Dobbs and Cohen (1980) report the results of isotherm tests performed in the laboratory in order to determine GAC adsorptive capacities for 1,2,4-trichlorobenzene. The isotherm constant is 157 mgL$^{1/n}$/gmg$^{1/n}$ while 0.31 is the isotherm slope (1/n). They also report the results of a pilot plant study conducted by the Cincinnati Water Works. The GAC pilot plant, operating with a 7- to 15-minute empty bed contact time (EBCT), reportedly removed 95 to 100% of 1,2,4-trichlorobenzene during the 4 months of operation.

- One study investigated the removal of chlorinated benzenes by a GAC column with microbial activity versus a glass bead control column with only bacterial growth (Bower and McCarty, 1982). Wastewater containing 10 mg/L of 1,2,4-trichlorobenzene was continuously applied to both columns (2.5 mm ID $\times$ 250 mm) under aerobic conditions and at a 60-minute EBCT for a period of 2 years. The GAC column, packed with 5.0 gm of Filtrasorb-100 mixed with 155 gm of glass beads removed approximately 97% of 1,2,4-trichlorobenzene. The control column removed 90% of the 1,2,4-trichlorobenzene. Effluent concentrations of chlorinated benzenes did not change significantly during the 2 years of operation.

- The efficiency of PAC in removing volatile organics, including 1,2,4-trichlorobenzene, was investigated at DuPont's Chambers Works Wastewater Treatment Plant (Hutton, 1981). Nuchar SA-15 activated carbon was used at a dosage of 114 mg/L for the treatment of an average 37 million gallons per day (MGD) wastewater containing 169 mg/L soluble total organic carbon (TOC). PAC was fed upstream of the aeration chamber designed for 8 hours aeration time. The concentration of 1,2,4-trichlorobenzene in the untreated wastewater is 210 μg/L. The results show a 99 + % removal efficiency of 1,2,4-trichlorobenzene by this process.

- Reinhard et al. (1986) report the ability of a full-scale RO treatment plant equipped with cellulose acetate (CA) membranes, and two pilot RO plants equipped with polyamide (PA) membranes and operated in parallel to remove TOC and a variety of trace organics, including 1,2,4-trichlorobenzene. The full-scale RO plant was operated at 460 psi, while the pilot-scale RO plants were operated at 250 psi. The full-scale RO plant removed an average of 56% of 1,2,4-trichlorobenzene from a mean feed concentration of 0.16 μg/L at a water recovery rate of 85%. The two pilot-scale RO plants removed 96% and 56% of 1,2,4-

trichlorobenzene from a mean feed concentration of 0.46 μg/L at water recovery rates of 54% and 67%, respectively.

- U.S. EPA (1986c) estimated the feasibility of removing 1,2,4-trichlorobenzene from water by packed column aeration, employing the engineering design procedure and cost model presented at the 1983 National ASCE Conference on Environmental Engineering. Based on chemical properties and assumed operating conditions, a 99% removal efficiency of 1,2,4-trichlorobenzene was postulated for a column with a diameter of 6.0 feet and packed with 28.3 feet of 1-inch plastic saddles. The air-to-water ratio required to achieve this degree of removal effectiveness is 1:27.

- Van Dyke et al. (1986) studied and reported the efficiency of a home-use water filter, containing pressed carbon block as filtering media, for the removal of a number of organic chemicals, including 1,2,4-trichlorobenzene. The filtering system consisted of a nonwoven prefilter, a pressed carbon block, and a porous polyethylene fritted core. The water was supplied at a constant pressure of 50 psig. Each run consisted of passing a volume of water equal to 150% of the filter-rated life of 500 gallons and analyzing for the various contaminants. 1,2,4-Trichlorobenzene was present in the influent at a concentration of 81 μg/L. This system removed 1,2,4-trichlorobenzene to below the detection limit (0.1 μg/L), a removal efficiency of greater than 99%.

- Lykins et al. (1986) reported the results of the effects of the four most commonly used disinfectants in the United States (chlorine, monochloramine, chlorine dioxide, and ozone) on the effluent from a water treatment plant. The effects of disinfection treatment by sand filtration and GAC adsorption on several organic compounds were evaluated in a pilot plant in Jefferson Parish, LA. 1,2,4-Trichlorobenzene was present in the nondisinfected influent at an average concentration of 6.2 ng/L. The disinfection system was designed for a contact time of 30 minutes. Chlorine, monochloramine and chlorine dioxide produced an increase in the total chlorobenzenes concentration. Ozone, however, reduced the total chlorobenzenes. Sand filtration reduced total chlorobenzenes by 26%. However, sand filtration effluent contained 43% more trichlorobenzenes than the nondisinfected influent. GAC reduced concentrations of total chlorobenzenes in the chlorinated stream by 73% and 43% in nondisinfected streams. No analytical data are presented on treated effluent composition.

VIII. References

ACGIH. 1983. American Conference of Governmental Industrial Hygienists. Supplement. Threshold limit values for chemical substances in work air adopted by ACGIH for 1982. Cincinnati, OH: ACGIH.

Ariyoshi, T., K. Ideguchi, K. Iwasaki and M. Arakaki. 1975a. Relationship between chemical structure and activity. II. Influences of isomers in dichlorobenzene, trichlorobenzene and tetrachlorobenzene on the activities of drug-metabolizing enzymes. Chem. Pharm. Bull. 23(4):824–830. (Cited in U.S. EPA, 1986a.)

Ariyoshi, T., K. Ideguchi, K. Iwasaki and M. Arakaki. 1975b. Relationship between chemical structure and activity. III. Dose-response or time-course of induction in microsomal enzymes following treatment with 1,2,4-trichlorobenzene. Chem. Pharm. Bull. 23(4):831–836. (Cited in U.S. EPA, 1986a.)

Black, W.D., V.E.O. Valli, J.A. Ruddick and D.C. Villeneuve. 1983. The toxicity of three trichlorobenzene isomers in pregnant rats. The Toxicologist. 3(1):30. Abstract.

Bower, E.J. and P.L. McCarty. 1982. Removal of trace chlorinated organic compounds by activated carbon on fixed-film bacteria. Environ. Sci. Technol. 16:836–843.

Brown, V.K.H., C. Muir and E. Thorpe. 1969. The acute toxicity and skin irritant properties of 1,2,4-trichlorobenzene. Ann. Occup. Hyg. 12:209–212 (Cited in U.S. EPA, 1986a.)

Carlson, G.P. 1977a. Halogenated benzenes, effect on xenobiotic metabolism and the toxicity of other chemicals. Ann. N.Y. Acad. Sci. 298:159–169.

Carlson, G.P. 1977b. Chlorinated benzene induction of hepatic porphyria. Experientia. 33(12):1627–1629.

Carlson, G.P. 1980. Effects of halogenated benzenes on arylesterase activity *in vivo* and *in vitro*. Res. Commun. Chem. Pathol. Pharmacol. 30(2):361–364.

Carlson, G.P. and R.G. Tardiff. 1976. Effect of chlorinated benzenes on the metabolism of foreign organic compounds. Toxicol. Appl. Pharmacol. 36:383–394.

Carlson, G.P., J.D. Dziezak and K.M. Johnson. 1979. Effect of halogenated benzenes on acetanilide esterase, acetanilide hydroxylase and procaine esterase in rats. Res. Commun. Chem. Pathol. Pharmacol. 5(1):181–184.

Coate, W.B., W.H. Schoenfisch, T.R. Lewis and W.M. Busey. 1977. Chronic inhalation exposure of rats, rabbits and monkeys to 1,2,4-trichlorobenzene. Arch. Environ. Health. 32(6):249–255.

Cragg, S.T., G.F. Wolfe and C.C. Smith. 1978. Toxicity of 1,2,4-trichlorobenzene in Rhesus monkeys: Comparison of two *in vivo* methods for estimating P-450 activity. Toxicol. Appl. Pharmacol. 5(1):340–341. Abstract.

Dobbs, R.A. and J.M. Cohen. 1980. Carbon adsorption isotherms for toxic organics. EPA 600/8-80-023. Cincinnati, OH: U.S. EPA Office of Research and Development, HERL, Wastewater Research Division.

Dow Chemical Company. 1958. Results of range-finding toxicological tests on 1,2,4-trichlorobenzene. Midland, MI: Biochemical Research Laboratory.

Girard, R., F. Tolot, P. Martin and J. Bourret. 1969. Serious blood disorders and exposure to chlorine derivatives of benzene (a report of 7 cases.) J. Med. Lyon. 50(1164):771–773. (Cited in U.S. EPA, 1986a.) (Fre.)

Hawley, G.G. 1981. Condensed Chemical Dictionary, 10th ed. New York, NY: Van Nostrand Reinhold Company.

Hutton, D.G. 1981. Removal of priority pollutants with a combined powdered activated carbon-activated sludge process. Chem. Water Reuse 2:403–428.

Jondorf, W.R., D.V. Parke and R.T. Williams. 1955. Studies in detoxication. 66. The metabolism of halogenobenzenes, 1,2,3-, 1,2,4- and 1,3,5-trichlorobenzenes. Biochem. J. 61:512–521. (Cited in U.S. EPA, 1986a.)

Kitchin, K.T. and M.T. Ebron. 1983. Maternal hepatic and embryonic effects of 1,2,4-trichlorobenzene in the rat. Environ. Res. 31:362–373.

Kociba, R.J., B.K. Leong and R.E. Hefner, Jr. 1981. Subchronic toxicity study of 1,2,4-trichlorobenzene in the rat, rabbit and beagle dog. Drug Chem. Toxicol. 4(3):229–249. (Cited in U.S. EPA, 1986a.)

Kohli, J., D. Jones and S. Safe. 1976. The metabolism of higher chlorinated benzene isomers. Can. J. Biochem. 54(3):203–208. (Cited in U.S. EPA, 1986a.)

Lawlor, T., S.R. Haworth and P. Voytek. 1979. Evaluation of the genetic activity of nine chlorinated phenols, seven chlorinated benzenes, and three chlorinated hexanes. Environ. Mutagen. 1:143. Abstract.

Lingg, R.D., W.H. Kaylor, S.M. Pyle, F.C. Kopfler, C.C. Smith, G.F. Wolfe and S. Cragg. 1982. Comparative metabolism of 1,2,4-trichlorobenzene in the rat and Rhesus monkey. Drug Metabol. Dispos. 10(2):134–141.

Love, O.T. Jr. and R.J. Miltner. 1985. Removal of volatile organic contaminants from ground water by adsorption. Cincinnati, OH: U.S. EPA Office of Research and Development.

Lykins, B.W., W.E. Koffskey and R.G. Miller. 1986. Chemical products and toxicologic effects of disinfection. J. AWWA. 7:(9):66–75.

Perry, R.H. and C.H. Chilton. 1973. Chemical Engineers Handbook, 5th ed. New York, NY: McGraw Hill Book Company.

Powers, M.B., W.B. Coate and T.R. Lewis. 1975. Repeated topical applications of 1,2,4-trichlorobenzene: effects on rabbit ears. Arch. Environ. Health. 30:165–167.

Reinhard, M., N.L. Goodman, P.L. McCarty and D.G. Arge. 1986. Removing trace organics by reverse osmosis using cellulose acetate and polyamide membranes. J. AWWA. 78(4):163–174.

Rimington, C. and G. Ziegler. 1963. Experimental porphyria in rats induced by chlorinated benzenes. Biochem. Pharmacol. 12:1387–1397. (Cited in U.S. EPA, 1986a.)

Robinson, K.S., R.J. Kavlock, N. Chernoff and L.E. Gray. 1981. Multigeneration study of 1,2,4-trichlorobenzene in rats. J. Toxicol. Environ. Health. 8(3):489–500.

Rowe, V.K. 1975. Written communication (cited in U.S. EPA, 1980.)

Ruddick, J.A., W.D. Black, D.C. Villeneuve and V.E. Valli. 1983. A teratological evaluation following oral administration of trichloro- and dichlorobenzene isomers to the rat. Teratology. 27(2):73A–74A. Abstract.

Schoeny, R.S., C.C. Smith and J.C. Loper. 1979. Non-mutagenicity for Salmonella of the chlorinated hydrocarbons arochlor 1254, 1,2,4-trichlorobenzene, mirex and kepone. Mutat. Res. 68(2):125–132.

Sittig, M. 1985. Handbook of Toxic and Hazardous Chemicals and Carcinogens, 2nd ed. Park Ridge, NJ: Noyes Publications.

Smith, E.N. and G.P. Carlson. 1980. Various pharmacokinetic parameters in relation to enzyme-inducing abilities of 1,2,4-trichlorobenzene and 1,2,4-tribromobenzene. J. Toxicol. Environ. Health. 6(4):737–749.

Smith, C.C., S.T. Cragg and G.F. Wolfe. 1978. Subacute toxicity of 1,2,4-trichlorobenzene (TCB) in subhuman primates. Fed. Proc. 37(3):248.

U.S. EPA. 1979. U.S. Environmental Protection Agency. Water-related environmental fate of 129 priority pollutants. Vol. II. EPA 440/4–79–029b. Washington, DC: U.S. EPA.

U.S. EPA. 1980. U.S. Environmental Protection Agency. Ambient water quality criteria for chlorinated benzenes. EPA 440/5–80–028. NTIS PB 81–117392. Cincinnati, OH: Environmental Criteria and Assessment Office.

U.S. EPA. 1984a. U.S. Environmental Protection Agency. Chlorinated hydrocarbons. Fed. Reg. 49(209):128–135. October 26.

U.S. EPA. 1984b. U.S. Environmental Protection Agency. Base/neutral and acids. Fed. Reg. 49(209):153–174. October 26.

U.S. EPA. 1986a. U.S. Environmental Protection Agency. Drinking water criterial document for trichlorobenzenes. External review draft. Cincinnati, OH: Environmental Criteria and Assessment Office.

U.S. EPA. 1986b. U.S. Environmental Protection Agency. Guidelines for carcinogenic risk. Fed. Reg. 51(185):33992–34003. September 24.

U.S. EPA. 1986c. U.S. Environmental Protection Agency. Evaluation of 1,2,4-trichlorobenzene removal from water by packed column air stripping. Prepared by Office of Water for Health Advisory Treatment Summaries.

U.S. EPA. 1988. U.S. Environmental Protection Agency. Occurrence and exposure assessment of trichlorobenzene in public drinking water supplies. Preliminary draft. Washington, DC: Office of Drinking Water.

Van Dyke, K., R. Kvennen, J. Stiles, J. Wezeman and J. O'Neal. 1986. Test stand design and testing for pressed carbon block water filter. Am. Lab. 18(9):118–132.

Verschueren, K. 1977. Handbook of Environmental Data on Organic Chemicals. New York, NY: Von Nostrand Reinhold Co.

Watanabe, P.G., R.J. Kociba, R.E. Hefner, Jr., H.O. Yakel and B.K.J. Leong. 1978. Subchronic toxicity studies of 1,2,4-trichlorobenzene in experimental animals. Toxicol. Appl. Pharmacol. 45(1):332–333. Abstract.

Yamamoto, H., Y. Ohno, K. Nakamori, T. Okuyama, S. Imai and Y. Tsubura. 1982. Chronic toxicity and carcinogenicity test of 1,2,4-trichlorobenzene on mice by dermal paintings. J. Nara. Med. Assoc. 33:132–145. (Cited in U.S. EPA, 1986a.) (Jap.)

Yang, K.H., R.E. Peterson and J.M. Fujimoto. 1979. Increased bile duct-pancreatic fluid flow in benzene and halogenated benzene-treated rats. Toxicol. Appl. Pharmacol. 47(3):505–514.

Bis-(2-Chloroisopropyl) Ether

I. General Information and Properties

A. CAS No. 108–60–1

B. Structural Formula

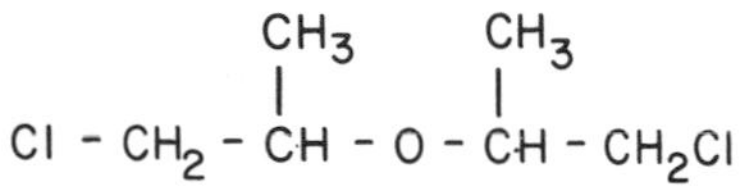

$$Cl - CH_2 - \overset{\overset{\displaystyle CH_3}{|}}{CH} - O - \overset{\overset{\displaystyle CH_3}{|}}{CH} - CH_2Cl$$

Bis-(2-chloroisopropyl) ether

C. Synonyms

- Bis-(2-chloro-1-methyl ethyl) ether; (2-chloro-1-methyl ethyl) ether; dichlorodiisopropyl ether; 2,2-dichloroisopropyl ether.

D. Uses

- Bis-(2-chloroisopropyl) ether is used as a solvent for fats, waxes, and greases; in textile manufacturing; in the manufacturing of cleaning solutions and spotting agents; in paint and varnish removers and as an intermediate in chemical synthesis (Verschueren, 1983).

E. Properties (U.S. EPA, 1980; Verschueren, 1983; Sax, 1984)

Chemical Formula	$C_6H_{12}Cl_2O$
Molecular Weight	171.07
Physical State	Colorless liquid
Boiling Point	189°C
Melting Point	-97 to -102°C
Density	—
Vapor Pressure	0.85 mm Hg
Specific Gravity	1.11
Water Solubility	1,700 mg/L
Log Octanol/Water Partition Coefficient	2.58

157

Taste Threshold (water)	—
Odor Threshold (water)	0.2–0.32 mg/L
Odor Threshold (air)	0.32 ppm
Conversion Factor	1 mg/m^3 = 0.143 ppm
	1 ppm = 7.0 mg/m^3

F. Occurrence

- Bis-(2-chloroisopropyl) ether has been detected in 0.1% of U.S. surface waters (Staples et al., 1985). It has also been detected in a few drinking water supplies. A concentration of 0.18 μg/L was detected in water from New Orleans (Keith et al., 1976). In a Report to Congress in 1975, U.S. EPA summarized that the highest concentration of bis-(2-chloroisopropyl) ether yet reported was 1.58 μg/L in drinking water (U.S. EPA, 1980).

G. Environmental Fate

- Bis-(2-chloroisopropyl) ether is expected to persist in water (U.S. EPA, 1980).

II. Pharmacokinetics

A. Absorption

- Bis-(2-chloroisopropyl) ether is readily absorbed after oral administration.

- Groups of three female CD rats were given single oral doses of 0.2 μg/kg, 3, 30 or 300 mg/kg ^{14}C-bis-(2-chloroisopropyl) ether labeled with ^{14}C at the β position (Smith et al., 1977). The compound was administered either in corn oil or in Emulphor. There was evidence of saturation kinetics at the highest dose but peak concentrations of radioactivity in the blood occurred 2 to 4 hours after dosing at the lower doses. Radioactivity was detected in the blood after 15 minutes, the earliest sampling time.

- Smith et al. (1977) also studied the absorption of bis-(2-chloroisopropyl) ether in monkeys. Two female rhesus monkeys that were given single oral doses of 30 mg/kg ^{14}C-bis-(2-chloroisopropyl) ether had maximum concentrations of radioactivity in the blood after 2 hours. Comparing the blood levels of ^{14}C-bis-(2-chloroisopropyl) ether in rats and monkeys indicates that monkeys have much higher blood levels than do rats during the first 8 hours postexposure.

B. Distribution

- Species differences were observed in the tissue distribution of bis-(2-chloroisopropyl) ether.

- Tissue distribution of radioactivity was determined 7 days after single 30 mg/kg injections of ^{14}C-bis-(2-chloroisopropyl) ether in three female CD rats intraperitoneal (i.p.) and one monkey intravenous (i.v.) (Smith et al., 1977). Distribution was widespread and the tissue concentrations were observed to vary between the rats and the monkeys. These varying tissue amounts, reported as percentages of administered radioactivity, were 1.98 for fat, 1.23 for muscle, 0.77 for blood, 9.36 for liver, 0.66 for kidneys and 0.03 for brain in rats; and 0.78 for fat, 2.76 for muscle, 0.74 for blood, 1.34 for liver, 0.05 for kidneys and 0.18 for brain in the monkey.

C. Metabolism

- 1-Chloro-2-propanol (0.1 to 1.0% of the administered dose), propylene oxide (not quantitated) and 2-(1-methyl-2-chloroethoxy) propionic acid (not quantitated) were identified as urinary metabolites in female CD rats treated with 30 mg/kg bis-(2-chloroisopropyl) ether by i.p. injection (Smith et al., 1977).

D. Excretion

- Urinary excretion of radioactivity was essentially complete 24 hours following i.p. (three rats) or i.v. (one monkey) injection of a single 30 mg/kg dose of ^{14}C-bis-(2-chloroisopropyl) ether, although the rats excreted approximately twice as much as the monkey (55 vs. 25% of the administered dose at 24 hours and 63 vs. 29% of the administered dose at 168 hours) (Smith et al., 1977). Approximately 1% and 6% of the administered radioactivity were eliminated by the monkey and rats, respectively, in the feces by 7 days.

III. Health Effects

A. Humans

- Information regarding effects of short-term or long-term exposure to bis-(2-chloroisopropyl) ether in humans could not be located in the available literature.

B. Animals

1. Short-term Exposure

- Oral LD_{50} in rats is 240 mg/kg (Smyth et al., 1951).

- Groups of five B6C3F$_1$ mice/sex were administered daily 17.8, 31.6, 56.2, 100, 187, 316 or 562 mg/kg doses of bis-(2-chloroisopropyl) ether in corn oil by gavage for 14 consecutive days (NTP, 1982). The test compound was contaminated with bis-(2-chloropropyl) ether and 2-chloro-1-methylethyl(2-chloropropyl) ether. The mice were observed daily for mortality, weighed on days 0, 7 and 14, and necropsied after 14 days. All mice of both sexes receiving the highest dose (562 mg/kg) died, indicating that the effect was treatment-related; one female at the 316 mg/kg dose, one female at the 100 mg/kg dose and one male at the 56.2 mg/kg dose also died. The mice receiving the highest dose had a hunched appearance, but other signs of toxicity, including gross lesions, were not observed at this or any of the other dose levels. No controls were used.

2. Dermal/Ocular Effects

- Data regarding dermal or ocular effects by bis-(2-chloroisopropyl) ether could not be located in the available literature.

3. Long-term Exposure

- Groups of 10 B6C3F$_1$ mice/sex were administered daily 0, 10, 25, 50, 100 or 250 mg/kg doses of bis-(2-chloroisopropyl) ether in corn oil by gavage, 7 days/week for 13 weeks (NTP, 1982). The isopropyl/n-propyl mixture used in the NTP (1982) 14-day study was used. The mice were observed daily for mortality and morbidity, observed weekly for clinical signs, and weighed weekly. Necropsies were performed on all animals, and comprehensive histological examinations were performed on all control and high-dose animals. Respiratory lesions (focal pneumonitis) at the three highest doses were the only effect reported. This study established a NOAEL of 25 mg/kg.

- Groups of 10 Fischer 344 rats/sex were given daily gavage doses of 0, 10, 25, 50, 100 or 250 mg/kg bis-(2-chloroisopropyl) ether in corn oil 7 days/week for 13 weeks (NCI, 1979). The isopropyl/n-propyl mixture as used in the NTP (1982) mouse studies described above was also used in this study. Terminal mean body weights were decreased 20% in the high-dose males, but treatment had no effect on survival or gross or histopathological findings in either sex at any dose. This study established a NOAEL of 100 mg/kg.

- Groups of 50 Fischer 344 rats/sex were treated by gavage with 70% bis-(2-chloroisopropyl) ether containing 2.1% bis-(2-chloro-n-propyl) ether and 28.5% 2-chloro-1-isopropyl (2-chloropropyl) ether in corn oil (NCI, 1979). Doses of 0, 100 or 200 mg/kg/day were administered 5 days/week for 103 to 105 weeks, and the animals were observed for 1 to 2 weeks. Toxicity endpoints were the same as those routinely assessed in NCI/NTP bioassays. Effects included dose-related decreased weight gain or weight loss and significantly decreased survival at the high dose in males, and high and low doses in females. Nonneoplastic pathological effects occurred at the high dose in both sexes and included centrilobular necrosis of the liver (22 vs. 10% in vehicle controls in males, and 15 vs. 2% in females) and esophageal hyperkeratosis. Treatment-related pathological effects were not observed at 100 mg/kg/day.

- Groups of five B6C3F$_1$ mice/sex were treated by gavage with 0, 100 or 200 mg/kg/day bis-(2-chloroisopropyl) ether in corn oil 5 days/week for 103 weeks and observed for 1 to 7 weeks (NTP, 1982). A 70% isopropyl isomer mixture essentially the same as that used in the NCI (1979) study in rats was tested. Toxicity endpoints were the same as those routinely assessed in the NCI/NTP bioassays. There were no treatment-related effects on clinical signs, body weight or survival. Chronic inflammation of the nasal cavity and/or nasolacrimal duct (0, 0 and 60% in the control, low-dose and high-dose groups, respectively) and fatty metamorphosis of the liver (2, 12 and 28% in the control, low-dose and high-dose groups, respectively) in the males were the only dose-related nonneoplastic effects.

- Bis-(2-chloroisopropyl) ether (98.5% pure) was administered in the diet to groups of 56 SPF-ICR mice/sex at concentrations of 0, 80, 400, 2,000 and 10,000 ppm (Mitsumori et al., 1979). Mice were killed as follows: 7/sex at weeks 13, 26 and 52; 6/sex at week 78 and the remaining mice at week 104. Body weights and food consumption were measured monthly throughout the study, and comprehensive hematological, blood biochemical and urinalysis determinations were performed at the interim and terminal sacrifices. Comprehensive gross and histopathological examinations were conducted on all animals at the scheduled sacrifices and on those that were moribund or died. Increased mortality, attributed to decreased food consumption and subsequent starvation, was evident at 10,000 ppm in both sexes. Other effects at 10,000 ppm were mild anemia (decreased erythrocyte count and/or hemoglobin concentration), increased polychromatic erythrocytes and splenic hemosiderin deposition in both sexes at weeks 13, 26 and/or 52. Of these effects, only hemosiderin deposition was evident at 104 weeks (34/112 treated mice of both sexes vs. 3/112 controls). Extramedullary hematopoiesis of the spleen was seen in the 10,000 ppm males at week 13 (the time of most pronounced anemia). At 2,000 ppm, similar treatment-related effects only occurred in the females (decreased body weight, anemia, splenic hemosiderin deposition). Although the above effects primarily occurred during the first half of the study and were mild in degree, the consistent hematological and histological evidence was concluded to be indicative of treatment-related erythrocyte destruction. Based on these effects, 2,000 ppm (198 mg/kg bw/day, determined by investigators) and 400 ppm (35.8 mg/kg bw/day) were considered to be the maximum NOAELs in the male and female mice, respectively.

4. Reproductive Effects

- Data regarding reproductive toxicity of bis-(2-chloroisopropyl) ether could not be located in the available literature.

5. Developmental Effects

- Data regarding developmental toxicity of bis-(2-chloroisopropyl) ether could not be located in the available literature.

6. Mutagenicity

- Bis-(2-chloroisopropyl) ether was mutagenic in *Salmonella typhimurium* strain TA100 when tested with or without S-9 metabolic activation preparation in suspension or as a vapor (Simmon et al., 1977).

- Bis-(2-chloroisopropyl) ether caused chromosome aberrations and sister-chromatid exchanges in cultured Chinese hamster ovary cells (NTP, 1982).

7. Carcinogenicity

- Bis-(2-chloroisopropyl) ether containing 2.1% bis-(2-chloro-n-propyl) ether and 28.5% 2-chloro-n-propyl ether in corn oil was administered by gavage to groups of 50 B6C3F$_1$ mice/sex at doses of 0, 100 or 200 mg/kg/day, 5 days a week for 103 weeks, followed by 1 to 7 weeks of observation (NTP, 1982). Treatment was carcinogenic, inducing increased incidences of alveolar/bronchiolar adenomas in both sexes and hepatocellular carcinomas in males. The occurrence of a low incidence of squamous cell papillomas or carcinomas in the stomach or forestomach in females (reportedly a rare tumor in B6C3F$_1$ mice) was considered to be treatment-related.

- Bis-(2-chloroisopropyl) ether containing 2.1% bis-(2-chloro-n-propyl) ether and 28.5% 2-chloro-n-propyl ether in corn oil was administered by gavage to groups of 50 Fischer 344 rats/sex at doses of 0, 100 or 200 mg/kg/day, 5 days/week for 103 to 105 weeks, followed by 1 to 2 weeks of observation (NCI, 1979). Treatment was not carcinogenic, but survival beyond week 75 was low in the high-dose groups.

- Bis-(2-chloroisopropyl) ether (98.5% pure) was administered in the diet to groups of 56 SPF-ICR mice/sex at concentrations of 0, 80, 400, 2,000 and 10,000 ppm (Mitsumori et al., 1979). Mice were killed as follows: 7/sex at weeks 13, 26 and 52; 6/sex at week 78 and the remaining mice at week 104. There were no significant differences between controls and treated mice in tumor incidences (comprehensive histological examinations were conducted), but survival beyond 78 weeks was low, particularly in the high-dose groups.

IV. Quantification of Toxicological Effects

A. One-day Health Advisory

Data were not sufficient for derivation of a One-day HA for bis-(2-chloroisopropyl) ether. The Longer-term HA for a 10-kg child of 4 mg/L (calculated below) is, therefore, adopted as a conservative One-day HA for a 10-kg child.

B. Ten-day Health Advisory

The study by NTP (1982), described under Short-term Exposure, was considered inadequate for the derivation of a Ten-day HA. This study used contaminated chemicals. Therefore, the Longer-term HA for a 10-kg child of 4 mg/L (calculated below) is recommended for the Ten-day HA.

C. Longer-term Health Advisory

The Mitsumori et al. (1979) study has been selected for evaluation as the basis for the Longer-term HA because mice (test species) seem to be more sensitive than the rats used in the only other alternative study (NCI, 1979). This study involved feeding SPF-ICR mice diets containing 0, 80, 400, 2,000 and 10,000 ppm of bis-(2-chloroisopropyl) ether (98.5% pure) for periods up to 104 weeks. At 13 weeks, 7 mice/sex were killed and evaluated. Anemia was seen in the 2,000 ppm female group and the 10,000 ppm male and female groups. A NOAEL of 2,000 ppm (198 mg/kg bw/day, as determined by the authors) for males and 400 ppm (35.8 mg/kg bw/day, as determined by the authors) for females can be determined at 13 weeks. In the more sensitive female mice, the Mitsumori et al. (1979) study provides the highest NOAEL of 35.8 mg/kg bw/day, which is below the dose levels reported for any observed adverse effect. Therefore, the NOAEL of 35.8 mg/kg bw/day is the most appropriate for calculation of the Longer-term HA. The study by NTP (1982) was considered inadequate for the derivation of HA values because it contained substantial quantities of impurities.

The Longer-term HA for a 10-kg child is calculated as follows:

$$\text{Longer-term HA} = \frac{(35.8 \text{ mg/kg bw/day}) (10 \text{ kg})}{(100) (1 \text{ L/day})} = 3.58 \text{ mg/L (rounded to 4,000 } \mu\text{g/L)}$$

The Longer-term HA for a 70-kg adult is calculated as follows:

$$\text{Longer-term HA} = \frac{(35.8 \text{ mg/kg bw/day}) (70 \text{ kg})}{(100) (2 \text{ L/day})} = 12.53 \text{ mg/L (rounded to 13,000 } \mu\text{g/L)}$$

D. Lifetime Health Advisory

The Mitsumori et al. (1979) chronic toxicity study has been selected to serve as the basis for the DWEL. In this study, groups of 56 SPF-ICR mice/sex were given diets containing bis-(2-chloroisopropyl) ether (98.5% pure) at concentrations of 0, 80, 400, 2,000 and 10,000 ppm. Subsets of the test mice were killed and evaluated as follows: 7/sex at weeks 13, 26 and 52; 6/sex at week 78; and the remaining mice at week 104. Increased mortality and adverse hematological effects were observed throughout the study in the 10,000 ppm groups. Hematological and histological evidence (anemia and splenic hemosiderin deposition) of treatment-related erythrocyte destruction was observed in the female 2,000 ppm group and male and female 10,000 ppm groups. The authors of the study determined the maximum NOAELs from their results to be 2,000 ppm (198 mg/kg bw/day, as determined by the authors) for the female mice. Since females seem to be more sensitive to bis-(2-chloroisopropyl) ether than males, the female NOAEL of 35.8 mg/kg bw/day will be used to derive the DWEL.

Step 1: Determination of the Reference Dose (RfD)

$$\text{RfD} = \frac{(35.8 \text{ mg/kg/day})}{(100) (10)} = 0.0358 \text{ mg/kg/day}$$

where:

10 = additional uncertainty factor for data gaps.

Step 2: Determination of the Drinking Water Equivalent Level (DWEL)

$$\text{DWEL} = \frac{(0.00358 \text{ mg/kg/day}) (70 \text{ kg})}{(2 \text{ L/day})} = 1.253 \text{ mg/L (rounded to 1,000 } \mu\text{g/L)}$$

Step 3: Determination of the Lifetime Health Advisory

Lifetime HA = (1.253 mg/L) (20%) = 0.250 mg/L (rounded to 300 μg/L)

E. Evaluation of Carcinogenic Potential

- In a 103-week gavage experiment, increased incidences of alveolar/ bronchiolar adenomas in both sexes and a low incidence of squamous cell papillomas or carcinomas in the forestomach (reportedly a rare tumor in B6C3F$_1$ mice) of females led the NTP (1982) to conclude that bis-(2-chloroisopropyl) ether was carcinogenic to mice. In addition, increased incidences of hepatocellular carcinomas were observed in treated males. Quantitative carcinogenic risk assessments for bis-(2-chloroisopropyl) ether could not be located in the available literature.

- Using the criteria described in U.S. EPA's guidelines for assessment of carcinogenic risk (U.S. EPA, 1986), bis-(2-chloroisopropyl) ether may be classified in Group D: not classifiable. This group is for substances with inadequate evidence of carcinogenicity.

- IARC has not evaluated the carcinogenic potential of bis-(2-chloroisopropyl) ether.

V. Other Criteria, Guidance and Standards

- Valid criteria, guidance or standards for bis-(2-chloroisopropyl) ether could not be located in the available literature.

VI. Analytical Methods

- Analysis of bis-(2-chloroisopropyl) ether is by a solvent extraction gas chromatographic procedure used for the determination of haloethers in water samples (U.S. EPA, 1984a). This method requires the extraction of a 1 L sample with methylene chloride using a separatory funnel. The methylene chloride extract is dried and exchanged to hexane during concentration to a volume of $\leq$ 10 mL. A portion of the concentration sample is injected into a gas chromatograph with a halogen specific detector. The reported method detection limit for bis-(2-chloroisopropyl) ether is 0.8 μg/L. Confirmatory analysis for this contaminant is by gas chromatography/mass spectrometry (U.S. EPA, 1984b). The detection limit for confirmation by mass spectrometry is 5.7 μg/L.

VII. Treatment Technologies

- Van Dyke et al. (1986) studied and reported the efficiency of a home-use water filter containing pressed carbon block as filtering media for the removal of a number of organic chemicals, including bis-(2-chloroisopropyl) ether. The filtering system consisted of a nonwoven prefilter, a pressed carbon block, and a porous polyethylene-fritted core. The water was supplied at a constant pressure of 50 psig. Each run consisted of passing a volume of water equal to 150% of the filter-rated life of 500 gallons and analyzing for the various contaminants. Bis-(2-chloroisopropyl) ether was present in the influent at a concentration of 105 μg/L. This system removed bis-(2-chloroisopropyl) ether to below its detection limit (0.2 μg/L).

- No data were found for the removal of bis-(2-chloroisopropyl) ether from drinking water by aeration. However, evaluation of the physical/chemical properties of bis-(2-chloroisopropyl) ether indicates that it may be amenable to removal by packed column aeration due to its relatively high vapor pressure.

VIII. References

Dow Chemical Co. n.d. Unpublished data. Biochemical Research Laboratory, Dow Chemical Co. (Cited in Kirwin and Sandmeyer, 1981.)

Fishbein, L. 1979. Potential halogenated industrial carcinogenic and mutagenic chemicals. III. Alkane halides, alkanols and ethers. Sci. Total Environ. 11:223–257.

Hawley, G.G. 1981. The Condensed Chemical Dictionary, 3rd ed. New York, NY: Van Nostrand Reinhold Co.

Keith, L.H., A.W. Garrison, F.R. Allen et al. 1976. Identification of organic compounds in drinking water from thirteen U.S. cities. In: Keith, L.H., ed. Identification and Analysis of Organic Pollutants in Water. Ann Arbor, MI: Ann Arbor Science. pp. 329–373.

Kirwin, C.J. and E.E. Sandmeyer. 1981. Ethers (Dichloroisopropyl Ether). In: Clayton, G.D. and F.E. Clayton, eds. Patty's Industrial Hygiene and Toxicology. 3rd ed., Vol. 2A. New York, NY: John Wiley and Sons, Inc. pp. 2519–2520.

Mitsumori, K., T. Usui, K. Takahashi and Y. Shirasu. 1979. Twenty-four month chronic toxicity studies of dichlorodiisopropyl ether in mice. Nippon NoYaku Gakkaishi. 4(3):323–335.

NCI. 1979. National Cancer Institute. Bioassay of technical grade bis(2-chloro-1-methylethyl) ether for possible carcinogenicity. NCI Carcinogenesis. Tech. Rep. Ser. No. 191. p. 84. [Also published as DHHS (NIH) 83-1795.]

NTP. 1982. National Toxicology Program. Carcinogenesis bioassay of bis-(2-chloro-1-methylethyl) ether (30%) in B6C2F$_1$ mice (gavage study). NCI Carcinogenesis. Tech. Rep. Serv. No. 239. p. 105. [Also published as DHHS (NIH) 83-1795.]

Perry, R.J. and C.H. Chilton. 1973. Chemical Engineer's Handbook. New York, NY: McGraw-Hill.

Sax, N.I., ed. 1984. Dangerous Properties of Industrial Materials, 6th ed. New York, NY: Van Nostrand Reinhold Co. 462 pp.

Simmon, V.F., K. Kauhanen and R.G. Tardiff. 1977. Mutagenic activity of chemicals identified in drinking water. Dev. Toxicol. Environ. Sci. 2:249-258.

Smith, C.C., R.D. Lingg and R.G. Tardiff. 1977. Comparative metabolism of haloethers. Ann. NY Acad. Sci. 298:111-123.

Smyth, H.F., C.P. Carpenter and C.S. Weil. 1951. Range-finding toxicity data. List IV. Arch. Ind. Hyg. Occup. Med. 4:119.

SRI. 1985. Stanford Research Institute. Directory of Chemical Producers. United States of America. Menlo Park, CA: SRI International. p. 448.

Staples, C.A., A.F. Werner and T.J. Hoogheem. 1985. Assessment of priority pollutant concentrations in the United States using STORET database. Environ. Toxicol. Chem. 4:131-142.

U.S. EPA. 1980. Ambient water quality criteria for chloralkyl ethers. EPA 440/5-80-030. NTIS PB 81-11741B. Prepared by the Office of Health and Environmental Assessment, Environmental Criteria and Assessment Office, Cincinnati, OH for the Office of Water Regulations and Standards, Washington, DC.

U.S. EPA. 1984a. U.S. EPA Method 611, Haloethers. *Fed. Reg.* 49(209):121-127.

U.S. EPA. 1984b. U.S. EPA Method 625, Base/Neutral and acids. *Fed. Reg.* 49(209):153-174.

U.S. EPA. 1986. Guidelines for carcinogen risk assessment. *Fed. Reg.* 51(185):33992-34003.

Van Dyke, K., R. Kuennen, J. Stiles, T. Wezeman and J. O'Neal. 1986. Test stand design and testing for a pressed carbon block water filter. American Laboratory. 18(9):118-132.

Verschueren, K. 1983. Handbook of Environmental Data on Organic Chemicals, 2nd ed. New York, NY: Van Nostrand Reinhold Co. p. 490.

p-Chlorotoluene

I. General Information and Properties

A. CAS No. 106–43–4

B. Structural Formula

p-Chlorotoluene

C. Synonyms

- 4-Chloro-1-methyl benzene, 1-chloro-4-methylbenzene, 4-chloro-toluene, p-tolyl chloride (U.S. EPA, 1985a; NIOSH, 1986).

D. Uses

- p-Chlorotoluene is used as a solvent and as a chemical intermediate in the manufacture of pesticides, dyestuffs, pharmaceuticals and peroxides (Gelfand, 1979).

E. Properties (Gelfand, 1979; Hawley, 1981; Valvani et al., 1981; Verschueren, 1983; U.S. EPA, 1985a; Ruth, 1986)

Chemical Formula	C_7H_7Cl
Molecular Weight	126.59
Physical State (at 25°C)	Colorless liquid
Boiling Point	162.4°C
Melting Point	6.5°C (Verschueren, 1983)
	7.5°C (Gelfand, 1979)
Density	—

169

Vapor Pressure (13.3°C)	96.6 mm Hg
Specific Gravity (25°C)	1.065 g/cm³
Water Solubility (20°C)	140 mg/L
Log Octanol/Water Partition Coefficient	3.33
Taste Threshold	—
Odor Threshold (water)	—
Odor Threshold (air)	0.2350 mg/m³ (unspecified isomer)
Conversion Factor	1 ppm = 5.17 mg/m³
	1 mg/m³ = 0.193 ppm

F. Occurrence

- p-Chlorotoluene was not detected in 73 whole water samples from ambient streams and wells for which records were contained in STORET (U.S. EPA, 1985a).

- Of 13 U.S. cities, only Miami, FL, contained detectable levels of p-chlorotoluene in ground water (1.5 μg/L) (Keith et al., 1976).

- Trace amounts of chlorotoluene (unspecified isomers) were found in 2/15 water samples collected in Buffalo and Niagara Falls, NY (Pellizzari et al., 1979). Elder et al. (1981) found the higher chlorotoluenes, but no monochlorotoluenes, in water and sediment samples collected at sites adjacent to hazardous waste disposal areas in Niagara Falls, NY.

- U.S. EPA (1985a) calculated a bioconcentration factor (BCF) of 230 for p-chlorotoluene in aquatic organisms, based on its K_{ow} values. No monitoring data for p-chlorotoluene in possible human food sources were available.

- p-Chlorotoluene was found in 95, 97 and 97% of the air samples collected in Newark, Elizabeth and Camden, NJ, with mean ambient air concentrations of 0.21, 0.25 and 0.22 ppb (1.09, 1.29 and 1.14 μg/m³), respectively (Harkov et al., 1983). In the Love Canal area of Niagara Falls, NY, Pellizzari et al. (1979) found isomers at concentrations of trace to 274 ng/m³ in 80% of the air samples collected. In the same area, Van Tassel et al. (1981) found chlorotoluene isomers in 50% of the air samples at concentrations of < 10 to 1,642 μg/m³. Concentrations of p-chlorotoluene from other unspecified areas of New York State ranged from 5 to 5.8 μg/m³ (Van Tassel et al., 1981). Pellizzari et al. (1979) also found chlorotoluene isomers at a concentration of 35 ng/m³ in 1/11 samples from Baton Rouge, LA.

G. Environmental Fate

- The Henry's Law constant calculated for p-chlorotoluene is 0.043 atm-m^3 M^{-1}. This value suggests that volatilization may be rapid from all types of surface waters (U.S. EPA, 1985a).

- No data concerning hydrolysis were available, but because chlorine is a ring substituent, hydrolysis is not expected to be an important process in determining the environmental fate of p-chlorotoluene (U.S. EPA, 1985a).

- Data concerning biodegradability of p-chlorotoluene are somewhat contradictory. Gibson et al. (1968) found that the compound was degraded by *Pseudomonas putida* that was grown in media with toluene as the sole carbon source. Gibson and Yeh (1973) found that the primary degradation process by *P. putida* was oxidation by the enzyme toluenedioxygenase, resulting in the degradation product, cis-3-methyl-6-chloro-3,5-cyclohexadiene-1,2-diol. In contrast, using a standardized biodegradation test that was developed for the Japanese Ministry of International Trade and Industry, Kawasaki (1980) and Sasaki (1978) classified this compound as degradation resistant. This test, however, often underestimates biodegradability, and therefore a negative result could be misleading (U.S. EPA, 1985a).

II. Pharmacokinetics

A. Absorption

- Data regarding absorption of p-chlorotoluene could not be located in the available literature. Toxic effects observed after oral administration of p-chlorotoluene indicate that gastrointestinal (GI) absorption occurs. Greater than 90% GI absorption has been observed with oral administration of o-chlorotoluene in rats (Wold and Emmerson, 1974; Quistad et al., 1983).

B. Distribution

- Data regarding distribution of p-chlorotoluene could not be located in the available literature. Gavage administration of o-chlorotoluene to rats was widely distributed, primarily in skin, lung, liver, kidney and fat (Quistad et al., 1983).

C. Metabolism

- p-Chlorotoluene administered to rabbits was excreted in urine as p-chlorobenzoic acid (Bray et al., 1955). The primary fecal and urinary metabolites of o-chlorotoluene after gavage administration to rats were β-glucuronides, mercapturic acid derivatives and amino acid conjugates (Quistad et al., 1983).

D. Excretion

- Data regarding excretion of p-chlorotoluene could not be located in the available literature. Gavage administration of o-chlorotoluene to rats was primarily excreted in the urine and, to a lesser extent, in feces and expired air (Quistad et al., 1983).

III. Health Effects

A. Humans

- Data regarding the toxicity of p-chlorotoluene to humans following short- or long-term exposure could not be located in the available literature.

B. Animals

1. Short-term Exposure

- Pis'ko et al. (1981) determined that the LD_{50} values for mice, rats and guinea pigs were 4,000, 5,500 and 3,750 mg/kg, respectively. Clinical signs included agitation followed by suppression, unsteady gait, disheveled fur, convulsive jerking, and rapid, superficial breathing.

- In a subacute 2-month experiment (Pis'ko et al., 1981) about 30 white rats/exposure group received daily intragastric doses of 55 or 550 mg/kg p-chlorotoluene in an oil solution. Unspecified dose-dependent effects on hemopoiesis, liver, kidneys, CNS and immune response were observed.

2. Dermal/Ocular Effects

- Pertinent data regarding the effects of dermal or ocular exposure to p-chlorotoluene could not be located in the available literature.

3. Long-term Exposure

- Pis'ko et al. (1981) administered daily, intragastric doses of 0.01, 0.1 and 1.0 mg/kg p-chlorotoluene in an oil solution to about 47 white rats/dose level for 6 months. The 0.1 and 1.0 mg/kg doses resulted in dose-related effects on hemopoiesis, neutrophil phagocytic activity, adrenal ascorbic acid content and liver enzymes. Other effects were noted in the liver, kidney, brain and central nervous system.

4. Reproductive Effects

- Reproductive function was examined in white rats (number not specified) treated with p-chlorotoluene in an oil solution (1) with a single dose of 1,100 or 1,833 mg/kg, (2) for 2 months with 55 or 550 mg/kg, or (3) for 6 months with 0.01, 0.1 or 1.0 mg/kg (time periods in relation to gestation not specified) (Pis'ko et al., 1981). A dose of 55 mg/kg for 2 months caused a significant increase in total embryonic mortality.

5. Developmental Effects

- Teratogenic and clastogenic effects of p-chlorotoluene were examined in 83 white rats (and 357 fetuses) treated (1) with a single dose of 1,100 or 1,833 mg/kg, (2) for 2 months with 55 or 550 mg/kg, or (3) for 6 months with 0.01, 0.1 or 1.0 mg/kg (time periods in relation to gestation not specified) (Pis'ko et al., 1981). In animals treated with a single dose of 1,833 mg/kg, only a tendency toward formation of single chromosome fragments was observed. p-Chlorotoluene was reported to produce no teratogenic and cytogenetic effects. The results from the other dosing regimens were not reported.

6. Mutagenicity

- p-Chlorotoluene was not mutagenic to *Salmonella typhimurium* (strains TA1535, TA1537, TA1538, TA98 and TA100) or to *Saccharomyces cerevisiae* (strain D3) with or without metabolic activation (Simmon et al., 1977).

7. Carcinogenicity

- Data regarding the carcinogenic effects of p-chlorotoluene could not be located in the available literature. The chemical has not been scheduled for testing by the National Toxicology Program (NTP, 1987).

IV. Quantification of Toxicological Effects

A. One-day Health Advisory

Data are insufficient for the derivation of a One-day HA for a 10-kg child. The Ten-day HA of 2 mg/L represents a conservative estimate for one-day exposure.

B. Ten-day Health Advisory

A Ten-day HA for p-chlorotoluene cannot be derived directly from compound specific data. Pis'ko et al. (1981) has not been considered for deriving a Ten-day HA due to the questionable significance of the reported observations. Since data currently available on p-chlorotoluene are insufficient for the derivation of a Ten-day HA for a 10-kg child, and since there is no evidence that the metabolism and/or toxicity of p-chlorotoluene differs significantly from o-chlorotoluene, the Ten-day HA derived for o-chlorotoluene has been adopted for use as the Ten-day HA for p-chlorotoluene.

Four groups of weanling Harlan rats (20 animals/sex) were given daily oral doses of 0, 20, 80, or 320 mg/kg o-chlorotoluene in gelatin capsules, containing an aqueous emulsion of o-chlorotoluene in 5% acacia (Gibson et al., 1974). A NOAEL of 20 mg/kg has been identified from this study. The Ten-day HA for the 10-kg child is calculated as follows:

$$\text{Ten-day HA} = \frac{(20 \text{ mg/kg/day}) (10 \text{ kg})}{(100) (1 \text{ L/day})} = 2 \text{ mg/L}$$

C. Longer-term Health Advisory

A Longer-term HA for p-chlorotoluene cannot be derived directly from compound specific data. Pis'ko et al. (1981) has not been considered for deriving a Longer-term HA due to the questionable significance of the reported observations. Since data currently available on p-chlorotoluene are insufficient for the derivation of a Longer-term HA for a 10-kg child, and since there is no evidence that the metabolism and/or toxicity of p-chlorotoluene differs significantly from o-chlorotoluene, the Longer-term HA derived for o-chlorotoluene has been adopted for use as the Longer-term HA for p-chlorotoluene.

Weanling Harlan rats (125 g each) were divided into four groups, each containing 20 animals/sex (Gibson et al., 1974). For 103 to 104 days, animals were administered by gavage 20, 80 or 320 mg/kg of o-chlorotoluene in an aqueous solution containing 5% acacia as the emulsifying agent. This study has been selected to serve as the basis for the Longer-term HA because an oral exposure route was used and rats were dosed for 103 to 104 days, an appropriate longer-term exposure period. The Longer-term HA for the 10-kg child is calculated as follows:

$$\text{Longer-term HA} = \frac{(20 \text{ mg/kg/day}) (10 \text{ kg})}{(100) (1 \text{ L/day})} = 2 \text{ mg/L } (2,000 \text{ }\mu\text{g/L})$$

The Longer-term HA for a 70-kg adult is calculated as follows:

$$\text{Longer-term HA} = \frac{(20 \text{ mg/kg/day}) (70 \text{ kg})}{(100) (2 \text{ L/day})} = 7 \text{ mg/L } (7,000 \text{ }\mu\text{g/L})$$

D. Lifetime Health Advisory

A Lifetime HA for p-chlorotoluene cannot be derived directly from compound specific data. Pis'ko et al. (1981) has not been considered for deriving a Lifetime HA due to the questionable significance of the reported observations. Since data currently available on p-chlorotoluene are insufficient for the derivation of a Lifetime HA, and since there is no evidence that the metabolism and/or toxicity of p-chlorotoluene differs significantly from o-chlorotoluene, the Lifetime HA derived for o-chlorotoluene has been adopted for use as the Lifetime HA for p-chlorotoluene.

A DWEL may be derived from the Gibson et al. (1974) subchronic oral exposure study by applying an additional uncertainty factor to account for a less than lifetime exposure.

Step 1: Determination of the Reference Dose (RfD)

$$\text{RfD} = \frac{20 \text{ mg/kg/day}}{1,000} = 0.02 \text{ mg/kg/day}$$

Step 2: Determination of the Drinking Water Equivalent Level (DWEL)

$$\text{DWEL} = \frac{(0.02 \text{ mg/kg/day}) (70 \text{ kg})}{2 \text{ L/day}} = 0.7 \text{ mg/kg/day } (700 \text{ }\mu\text{g/L})$$

Step 3: Determination of the Lifetime Health Advisory

$$\text{Lifetime HA} = (0.7 \text{ mg/L}) (20\%) = 0.14 \text{ mg/L (rounded to 0.1 mg/L or } 100 \text{ }\mu\text{g/L})$$

E. Evaluation of Carcinogenic Potential

- Data could not be located regarding the carcinogenicity of p-chlorotoluene. This chemical has not been scheduled for carcinogenicity testing by the National Toxicology Program (NTP, 1987).

- IARC has not evaluated the carcinogenic potential of p-chlorotoluene. Applying the criteria described in the U.S. EPA (1986a) guidelines for assessment of carcinogenic risk, p-chlorotoluene may be classified in

Group D: not classified. This category is for chemicals for which there is inadequate evidence of carcinogenicity from human and animal studies or no data are available.

V. Other Criteria, Guidance and Standards

- Data regarding criteria, guidelines or standards for p-chlorotoluene could not be located in the available literature. Pis'ko et al. (1981) recommended a maximum permissible concentration of 0.04 mg/L of both o- and p-chlorotoluene in water reservoirs.

VI. Analytical Methods

- Analysis of p-chlorotoluene is by a purge-and-trap gas chromatographic procedure used for the determination of volatile aromatic and unsaturated organic compounds in water (U.S. EPA, 1985b). This method calls for the bubbling of an inert gas through the sample and trapping volatile compounds on an adsorbent material. The adsorbent material is heated to drive off compounds onto a gas chromatographic column. The gas chromatograph is temperature programmed to separate the method analytes, which are then detected by the photoionization detector. Confirmatory analysis is by mass spectrometry (U.S. EPA, 1985c). The detection limit for this contaminant has not been determined for either method.

VII. Treatment Technologies

- The available information indicates that granular activated carbon (GAC) adsorption, resin adsorption and air stripping will remove p-chlorotoluene from contaminated water.

- Wood et al. (1980) report the results of a 2-year study at the Preston Water Treatment Plant in Florida. Two adsorbents were studied: Filtrasorb 400 as GAC adsorbent and ambersorb XE-340 and IRA-904 as resin adsorbents. Throughout the study, 1-inch diameter columns each packed with 176, 215 and 275 g of GAC, XE-340 and IRA-904, respectively, were used at a flow rate of 3 gpm/ft². The adsorbing units were connected as posttreatment to the conventional lime softening system. Raw water containing 0.02 to 0.2 μg/L p-chlorotoluene was treated by lime softening where an average 73% removal was reported. p-Chlorotoluene removal by GAC and resin adsorption was studied in the first phase of the program where raw water was fed directly to the adsorbing units. Both GAC and XE-340 removed all p-chlorotoluene, i.e., 100%, from an average concentration of 0.38 μg/L throughout the test period.

- U.S. EPA (1986b) estimated the feasibility of removing p-chlorotoluene from water by air stripping, employing the engineering design procedure and cost model presented at the 1983 National ASCE Conference on Environmental Engineering. Based on chemical and physical properties and assumed operating conditions, 99% removal efficiency of p-chlorotoluene was reported by a column with a diameter of 5.5 feet and packed with 21 feet of 1-inch plastic saddles. The air-to-water ratio required to achieve this degree of removal effectiveness is 1:14. Actual system performance data, however, are necessary to realistically determine the feasibility of using air stripping for the removal of p-chlorotoluene from contaminated drinking water supplies.

- In summary, a number of techniques for the removal of p-chlorotoluene from water have been examined. While the data are not unequivocal, it appears that GAC and resin adsorption are likely to be successful treatment techniques. The amenability of p-chlorotoluene to aeration has been clearly established. Selection of individual or combinations of technologies to attempt p-chlorotoluene removal from contaminated drinking water must be based on a case-by-case technical evaluation, and on assessment of the economics involved.

VIII. References

Bray, H.G., B.G. Humphris, W.V. Thorpe, K. White and P.B. Wood. 1955. Kinetic studies on the metabolism of foreign organic compounds. Biochem. J. 59:162–167.

Elder, V.A., B.L. Proctor and R.A. Hites. 1981. Organic compounds found near dump sites in Niagara Falls, New York. Environ. Sci. Technol. 15(10):1237–1243.

Gelfand, S. 1979. Chlorocarbons, hydrocarbons (toluene). In: Kirk, R.E., and D.F. Othmer, eds. Encyclopedia of Chemical Technology, Vol. 5, 3rd ed. New York, NY: John Wiley and Sons. pp. 819–827.

Gibson, D.T. and W.K. Yeh. 1973. Microbial degradation of aromatic hydrocarbons. In: The Microbial Degradation of Oil Pollutants. Center for Wetland Resources, LSU-SG-73-01. pp. 33–38. (Cited in U.S. EPA, 1985a.)

Gibson, D.T., J.R. Koch, C.L. Schuld and R.E. Kallio. 1968. Oxidative degradation of aromatic hydrocarbons by microorganisms. II. Metabolism of halogenated aromatic hydrocarbons. Biochemistry. 7(11):3795–3802.

Gibson, W.R., F.O. Gossett, G.R. Koenig and F. Marroquin. 1974. The toxicity of daily oral doses of o-chlorotoluene in the rat. Toxicology Division, Lilly Research Laboratories. Submitted to Test Rules Development Branch, Office of Toxic Substances, U.S. EPA, Washington, DC.

Harkov, R., B. Kebbekus, J.W. Bozzelli and P.J. Lioy. 1983. Measurement of selected volatile organic compounds at three locations in New Jersey during the summer season. J. Air Pollut. Control Assoc. 33(12):1177–1183.

Hawley, G.G. 1981. The Condensed Chemical Dictionary, 10th ed. New York, NY: Van Nostrand Reinhold Co. p. 243.

Kawasaki, M. 1980. Experiences with the test scheme under the chemical control law of Japan: an approach to structure-activity correlations. Ecotoxicol. Environ. Safety. 4:444–454.

Keith, L.W., A.W. Garrison, F.R. Allen et al. 1976. Identification of organic compounds in drinking water from thirteen U.S. cities. In: Keith, L.H., ed. Identification and Analysis of Organic Pollutants in Water. Ann Arbor, MI: Ann Arbor Science. pp. 329–373.

NIOSH. 1986. National Institute for Occupational Safety and Health. RTECS (Registry of Toxic Effects of Chemical Substances). March 1986: Online.

NTP. 1987. National Toxicology Program. Toxicology Research and Testing Program. Management Status Report. Research Triangle Park, NC: NTP. April 15.

Pellizzari, E.D., M.D. Erickson and R.A. Zweidinger. 1979. Formulation of preliminary assessment of halogenated organic compounds in man and environmental media. Research Triangle Park, NC: U.S. EPA. EPA 560/13-79-006.

Pis'ko, G.T., T.V. Tolstopyatova, T.V. Belyanina et al. 1981. Study of maximum permissible concentrations of o- and p-chlorotoluenes in bodies of water. Gig. Sanit. (8):67–68. (Rus. translation).

Quistad, G.B., K.M. Mulholland and G.C. Jamieson. 1983. 2-Chlorotoluene metabolism by rats. J. Agric. Food Chem. 31(6):1158–1162.

Ruth, J.H. 1986. Odor thresholds and irritation levels of several chemical substances: a review. Am. Ind. Hyg. Assoc. J. 47:A142-A151.

Sasaki, S. 1978. The scientific aspects of the chemical substances control law in Japan. In: Hutzinger, O., L.H. Van Letyoeld and B.C.J. Zoetman, eds. Aquatic Pollutants: Transformation and Biological Effects. Elmsford, NY: Pergamon Press. pp. 283–298.

Simmon, V.F., K. Kauhanen and R.G. Tardiff. 1977. Mutagenic activity of chemicals identified in drinking water. Dev. Toxicol. Environ. Sci. 2:249–258.

U.S. EPA. 1985a. U.S. Environmental Protection Agency. Health and environmental effects profile for chlorotoluenes (o-, m-, p-). Prepared by the Office of Health and Environmental Assessment, Environmental Criteria and Assessment Office, Cincinnati, OH, for the Office of Solid Waste, Washington, DC.

U.S. EPA. 1985b. U.S. Environmental Protection Agency. U.S. EPA Method 503.1–Volatile aromatic and unsaturated organic compounds in water by purge and trap gas chromatography. Environmental Monitoring and Support Laboratory, Cincinnati, OH. June 1985 (Revised November 1985).

U.S. EPA. 1985c. U.S. Environmental Protection Agency. U.S. EPA Method 524.1–Volatile organic compounds in water by purge and trap gas chromatography/mass spectrometry. Environmental Monitoring and Support Laboratory, Cincinnati, OH, June 1985. (Revised November 1985.)

U.S. EPA. 1986a. U.S. Environmental Protection Agency. Guidelines for carcinogen risk assessment. *Fed. Reg.* 51(185):33992–34003. September 24.

U.S. EPA. 1986b. U.S. Environmental Protection Agency. Economic evaluation of p-chlorotoluene removal from water by packed column air stripping. Prepared by Office of Water for Health Advisory Treatment Summaries, 1986.

Valvani, S.C., S.H. Yalkowsky and T.J. Roseman. 1981. Solubility and partitioning. IV. Aqueous solubility and octanol-water partition coefficients of liquid nonelectrolytes. J. Pharm. Sci. 70(5):502–507.

Van Tassel, S., N. Amalfitano and R.S. Narang. 1981. Determination of arenes and volatile haloorganic compounds in air at microgram per cubic meter levels by gas chromatography. Anal. Chem. 53:2130–2135.

Verschueren, K. 1983. Handbook of Environmental Data on Organic Chemicals, 2nd ed. New York, NY: Van Nostrand Reinhold Co. p. 387.

Wold, J.S. and J.L. Emmerson. 1974. The metabolism of ^{14}C-o-chlorotoluene in the rat. Pharmacologist. 16(2):196. Abstract.

Wood, P.R. et al. 1980. Removing potential organic carcinogens and precursors from drinking water. U.S. EPA, Office of Research and Development. EPA-600/2−8-130a.

1,2,3-Trichloropropane

I. General Information and Properties

A. CAS No. 96–18–4

B. Structural Formula

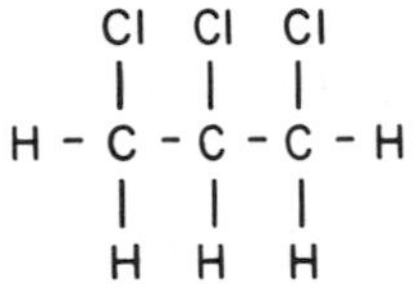

```
     Cl  Cl  Cl
     |   |   |
H  - C - C - C - H
     |   |   |
     H   H   H
```

1,2,3-Trichloropropane

C. Synonyms

- Allyl trichloride, glycerol trichlorohydrin, glyceryl trichlorohydrin, trichlorohydrin (NIOSH, 1986).

D. Uses

- 1,2,3-TCP is used as a paint and varnish remover, solvent, degreasing agent, and crosslinking agent in the clastomer Thiokiol ST (U.S. EPA, 1983).

E. Properties (Weast, 1972; Verschueren, 1977; Koneman, 1981; U.S. EPA, 1983)

Chemical Formula	$C_3H_5Cl_3$
Molecular Weight	147.43
Physical State (at 25°C)	Colorless clear liquid
Boiling Point	
(25 mm Hg)	156.85°C
Melting Point	–14.7°C
Density (20°C)	1.387
Vapor Pressure (20°C)	2 mm Hg
Specific Gravity (20°C)	1.3889

Water Solubility (20°C)	1,900 mg/L
Log Octanol/Water	
Partition Coefficient	2.63
Taste Threshold	—
Odor Threshold (water)	—
Odor Threshold (air)	—
Conversion Factor	1 ppm = 6 mg/m³

F. Occurrence

- Drinking water from Carrollton Water Plant in New Orleans, LA, contained < 0.2 µg/L of 1,2,3-TCP (Keith et al., 1976). It was also detected in Ames, IA, drinking water; however, levels were not given (U.S. EPA, 1976). Kool et al. (1982) reported that 1,2,3-TCP had been detected in drinking water in unspecified locations.

- Surface water from the Delaware River basin contained trichloropropane (unspecified isomer) at concentrations > 1 µg/L in 3% of samples (Dewalle and Chian, 1978). Wakeham et al. (1983) found trichloropropane in seawater of Narragansett Bay, RI, but concentrations were not reported.

G. Environmental Fate

- Dilling (1977) reported that the half-life for evaporation of 1,2,3-TCP from water was about 1 hour under the following conditions: 0.92 ppm aqueous solution; 6.5 cm deep; 200 rpm stirring; 25°C; < 0.2 mph air current.

- Matsui et al. (1975) determined that trichloropropane (unspecified isomer) was relatively easy to decompose by microbes in activated sludge.

II. Pharmacokinetics

A. Absorption

- Data regarding the absorption of 1,2,3-TCP could not be located in the available literature.

B. Distribution

- Volp et al. (1984) administered 3.6 mg/kg bw C¹⁴ 1,2,3-TCP (label at 1,3 carbon) intravenously to male Fischer 344 rats. Tissues and excreta were analyzed for total radioactivity and unchanged 1,2,3-TCP at various time periods following the administration of 1,2,3-TCP. The distribution and excretion of 1,2,3-TCP were rapid. 37% of the dose was accounted for in adipose tissue within 15 min. This consisted of primarily unchanged 1,2,3-TCP. The largest fraction of the dose was detected in the liver in the form of metabolites after a 4-hour exposure of 1,2,3-TCP. The kidneys also accumulated radiolabeled 1,2,3-TCP with a peak of 2.8% of the total dose at 2 hours, decreasing thereafter to < 1% at the end of the 24-hour period. The small intestine had a concentration of 9.3% of the dose at 1 hour. Brain, lungs, spleen, testes and epididymidis contained < 0.5% of the total dose at all times. The primary sites of distribution associated with radiolabeled 1,2,3-TCP were initially the adipose tissue, skin and muscle, then subsequently the liver.

C. Metabolism

- In the Volp et al. (1984) study described above, 1,2,3-TCP was rapidly distributed to all tissues, specifically adipose tissue, skin and muscle. After 4 hours, the concentration of unchanged 1,2,3-TCP was 90% of the total radiolabeled compound in adipose tissue (3.8% of the dose). This concentration was observed to decrease to 37% at the 24-hour period following the administration of compound. The investigators reported that (1) the major metabolite of 1,2,3-TCP was carbon dioxide (25% of dose), and (2) other minor metabolites were also present but not identified.

D. Excretion

- In the Volp et al. (1984) study described above 99% of the dose was excreted within 6 days. Most of the excretion (90%) occurred in the first 24-hours, with urine being the principal route. Of the total dose of radioactivity, 40% was excreted in urine, 30% in expired air, and 18% in the feces in the first 24 hours. The urine contained no detectable 1,2,3-TCP, indicating that all of the radiolabel was 1,2,3-TCP metabolites. Of the 30% of the initial dose of radioactivity eliminated in expired air, 5% was unchanged TCP and 25% was CO_2. Almost all of the unchanged TCP (85%) was expired within 30 minutes. In the bile, 30% of the total dose appeared within 6 hours, 5% of which was unchanged 1,2,3-TCP. The elimination half-time for unchanged 1,2,3-TCP was 30 to 45 hours for all major tissues. The half-time for elimination of radiolabel by all routes was 44 hours.

III. Health Effects

A. Humans

1. Short-term Exposure

- Silverman et al. (1946) exposed an average of 12 volunteers (males and females) to 1,2,3-TCP and other industrial solvent vapors for 15 minutes, and found that 100 ppm 1,2,3-TCP (600 mg/m^3) caused eye and throat irritation and had an unpleasant odor. A "borderline majority" of the subjects said that 50 ppm (300 mg/m^3) would be acceptable for an 8-hour workday. However, this level is based on organoleptic quality and not toxicity.

2. Long-term Exposure

- Pertinent data regarding long-term exposure of humans to 1,2,3-TCP could not be located in the available literature.

B. Animals

1. Short-term Exposure

- Saito-Suzuki et al. (1982) reported that 500 mg/kg bw 1,2,3-TCP by gastric intubation to male Sprague-Dawley rats was lethal. Smythe et al. (1962) reported an oral LD$_{50}$ of 0.32 mL/kg bw (444 mg/kg) 1,2,3-TCP for Carworth-Wistar male rats. An oral LD$_{50}$ of 320 mg/kg for 1,2,3-TCP also has been reported in the literature (RTECS, 1978).

- Wright and Schaffer (1932) administered one dose (route not specified) of 0.2 to 0.5 cc/kg (278–694 mg/kg) 1,2,3-TCP to three dogs. All dogs died within 1–2 days after dosing. The major signs of toxicity were narcosis and liver and kidney tissue necrosis.

- Shcherban and Piten'ko (1976) reported the following LD$_{50}$ values for 1,2,3-TCP (route not specified): rats, 505 mg/kg; mice, 369 mg/kg; rabbits, 380 mg/kg; and guinea pigs, 340 mg/kg. They also reported that 0.0035 mg/kg was a completely nontoxic dose. No other details were given.

- Several short-term inhalation studies using 1,2,3-TCP were available. Smythe et al. (1962) reported that 5/6 rats died when exposed to 6,000 mg/m^3 (1,000 ppm) for 4 hours. McOmie and Barnes (1949) reported that exposure to a vapor concentration of 30,000 mg/m^3 (5,000 ppm) for 20 minutes killed several mice (8/15 within 2 days). Four of the remaining seven mice died from liver damage 7 to 10 days later. When exposed to 15,000 mg/m^3 (2,500 ppm), 10 minutes/day for 10 days, 7/10 mice died. Lewis (1979) exposed rats and guinea pigs (five/sex) to 4,800 mg/ m^3 (799 ppm), 12,480 mg/m^3 (2,080 ppm) or 30,060 mg/m^3 (5,010 ppm) for 30 minutes, resulting in dose-related central nervous system (CNS) depression. Six guinea pigs and two rats at the high dose, 30,060 mg/m^3 (5,010 ppm), died.

- Sidorenko et al. (1979) exposed white male rats (strain not specified) to 2 to 800 mg/m^3 (0.33 to 133 ppm) 1,2,3-TCP for periods ranging from 2 hours to 86 days. An increase in the activity of blood catalase, acetylcholinesterase, and the excitability of nerve centers was reported. These changes were observed in rats after 4 hours of exposure to 800 mg/m^3 (133 ppm) and after 40 days of exposure to 2 mg/m^3 (0.33 ppm).

2. Dermal/Ocular Effects

- Smythe et al. (1962) reported a single skin penetration LD_{50} of 1.77 mL/ kg (2,458 mg/kg) 1,2,3-TCP for rabbits. On a scale of 1 to 10 (1 = least severe, 10 = most severe), 1,2,3-TCP rated 1 for skin irritation and 4 for corneal injury.

- McOmie and Barnes (1949) determined 1,2,3-TCP to be an "intense skin irritant" for rabbits, partially due to its lipid solvent properties. Repeated applications led to sloughing and cracking preceded by irritation and erythema. In a 15-day period, 10 applications of 2 mL/100 cm² led to pain, subdermal hemorrhage and death in 1/7 treated rabbits. The other six rabbits survived and healed within 6 weeks.

3. Long-term Exposure

- 1,2,3-TCP was administered by gavage in corn oil, 5 days/week for 120 days, to Fischer 344 rats (20/sex/group) at dose levels of 8, 16, 32, 63, 125 and 250 mg/kg bw/day (NTP, 1983a). One control group of 30 rats/ sex received corn oil. All animals in the 250 mg/kg dose group died as a result of treatment, with the main findings being renal and hepatic toxicity and necrosis and inflammation of the nasal mucosa. Mortality was also observed in the 125 mg/kg dose group. Dose-related clinical effects (e.g., thin and hunched appearance, depression, abnormal eyes and urine stains) were observed in female rats at doses of ≥125 mg/kg. Hematological effects (decreased hematocrit, hemoglobin and erythrocyte counts) were seen in both sexes at doses of ≥16 mg/kg. There was a dose-related increase in liver and kidney weights. At 125 mg/kg there was also an increase in the weight of testes and a decrease in epididymis weight, but no histomorphologic change was observed. Principal target organs were the liver and kidney, with histomorphological and clinical chemistry changes observed in dose groups ≥63 mg/kg. The nasal turbinates were also a target, but it was suggested that this may have been due to a local effect as opposed to a systemic effect. The NOAEL for this study is 8 mg/kg and the LOAEL is 16 mg/kg based on hematologic effects.

- 1,2,3-TCP was administered by gavage in corn oil, 5 days/week for 120 days, to B6C3F$_1$ mice (20/sex/group) at dose levels of 8, 16, 32, 63, 125 and 250 mg/kg bw/day (NTP, 1983b). One control group of 30 mice/sex received corn oil. Treatment-related deaths due primarily to hepatic toxicity occurred particularly in males at the 250 mg/kg level. The principal target organs were the liver, lung, kidney and stomach, with effects also seen in the spleen and nasal passages. Body weight gain was not affected except for a decrease in two male survivors in the 250 mg/kg group. Evaluation of the hematological and clinical chemistry data revealed no changes of biological importance since findings were sporadic in distribution and noted by the authors as "incidental to compound administration." Increased weights/or ratios were noted in the liver and thymus at doses ≥ 125 mg/kg. The lowest dose with a statistically significant effect was 16 mg/kg, which resulted in a lower brain weight ratio in female mice. This is the basis for defining 16 mg/kg as the LOAEL for this study. The NOAEL is 8 mg/kg.

4. Reproductive Effects

- Johannsen et al. (1988) reported the results of reproduction studies in the rat following repeated inhalation exposure. Groups of 10 male and 20 female rats were exposed 6 hours/day, 5 days/week to 5 ppm (30 mg/ m^3) or 15 ppm (90 mg/m^3) 1,2,3-trichloropropane vapor during premating and mating. Female rats were also exposed during gestation. Investigators stated that (1) the body weights of both sexes of the 5 ppm (30 mg/m^3) level were comparable to control values, (2) at the 15 ppm (90 mg/m^3) level, both sexes exhibited lower mean body weights and significantly (p ≤ 0.01) lower mean weight gains during the premating period, (3) mating performance was low in all groups of female rats including the controls, and (4) all measured progeny indices appeared unaffected by inhalation exposure of 1,2,3-trichloropropane.

5. Developmental Effects

- No treatment-related effects on incidence of grossly visible internal or external malformations occurred in the offspring of female Sprague-Dawley rats injected intraperitoneally with 37 mg/kg bw 1,2,3-TCP in corn oil on days 1 through 15 of gestation (Hardin et al., 1981).

- Hardin et al. (1981) administered by intraperitoneal (i.p.) injection 37 mg/kg bw 1,2,3-TCP in corn oil to groups of 10 to 15 pregnant Sprague-Dawley rats on days 1 through 15 of gestation. Exposure caused maternal toxicity as indicated by reduced body weight gain or altered organ weights in two or more organs, but did not cause fetotoxicity (reduced fetal size or reduced survival rate).

6. Mutagenicity

- Stolzenberg and Hine (1980) reported that 1,2,3-TCP was mutagenic to *Salmonella typhimurium* only with a microsomal activating system (S-9). At the two concentrations evaluated on the tester strain TA100, a dose-dependent increase in revertant colony numbers was observed in the presence of S-9 mix but not in its absence.

- Results were negative in a dominant lethal assay in which 80 mg/kg bw/day 1,2,3-TCP dissolved in olive oil was administered by gastric intubation to Sprague-Dawley rats for 5 consecutive days (Saito-Suzuki et al., 1982).

- In a dominant lethal assay in which 15 male Sprague-Dawley rats received gavage doses of 80 mg/kg bw/day for 5 consecutive days prior to mating, no effects were seen on reproductive performance (frequency of fertile matings) (Saito-Suzuki et al., 1982). No testicular lesions were observed.

7. Carcinogenicity

- Pertinent data regarding the carcinogenicity of 1,2,3-TCP could not be located at the time of this publication. The National Toxicology Program is currently conducting a 2-year gavage study in rats and mice (NTP, 1988). A judgment of carcinogenicity will be deferred until this study is completed.

IV. Quantification of Toxicological Effects

A. One-day Health Advisory

Sufficient data are not available for the derivation of a One-day HA for 1,2,3-TCP. Available oral data in rats (Saito-Suzuki et al., 1982; Smythe et al., 1962) and dogs (Wright and Schaffer, 1932) define lethal dosages, but sublethal effects were not investigated. In absence of toxicity data, the Longer-term HA value for a child (600 µg/L) is recommended at this time.

B. Ten-day Health Advisory

Sufficient data are not available for the derivation of a Ten-day HA for 1,2,3-TCP. Several Russian inhalation studies (Sidorenko et al., 1979; Belyaeva et al., 1977; Tarasova, 1975) reported that adverse effects occurred in rats exposed to concentrations as low as 2 mg/m^3, but exposure schedules were not provided and these studies were not available for review. In absence of toxicity data, the Longer-term HA value for a child (600 μg/L) is recommended at this time.

C. Longer-term Health Advisory

The NTP studies (1983a,b) have been chosen to serve as the basis for the Longer-term HA. Fischer 344 rats and B6C3F$_1$ mice were administered 1,2,3-TCP by gavage 5 days/week for 120 days. For both rats and mice, the lowest dose level of 8 mg/kg was a NOAEL, while 16 mg/kg was a LOAEL based on hematological effects in the rats and brain weight changes in the mice.

The Longer-term HA for a 10-kg child is calculated as follows:

$$\text{Longer-term HA} = \frac{(8 \text{ mg/kg/day}) (5/7) (10 \text{ kg})}{(100) (1 \text{ L/day})} = 0.57 \text{ mg/L} (600 \text{ } \mu\text{g/L})$$

where:

$5/7$ = factor to account for exposure of 5 out of 7 days.

The Longer-term HA for a 70-kg adult is calculated as follows:

$$\text{Longer-term HA} = \frac{(8 \text{ mg/kg/day}) (5/7) (70 \text{ kg})}{(100) (2 \text{ L/day})} = 2 \text{ mg/L} (2,000 \text{ } \mu\text{g/L})$$

where:

$5/7$ = factor to account for exposure of 5 out of 7 days.

D. Lifetime Health Advisory

There are no studies of suitable duration for the derivation of a DWEL. Therefore, the subchronic studies by NTP (1983a,b) will be used.

Step 1: Determination of the Reference Dose (RfD)

$$\text{RfD} = \frac{(8 \text{ mg/kg/day}) (5/7)}{(1,000)} = 0.006 \text{ mg/kg/day}$$

where:

$5/7$ = factor to account for exposure of 5 out of 7 days.

Step 2: Determination of the Drinking Water Equivalent Level (DWEL)

$$\text{DWEL} = \frac{(0.006 \text{ mg/kg/day}) (70 \text{ kg})}{(2 \text{ L/day})} = 0.2 \text{ mg/L } (200 \text{ }\mu\text{g/L})$$

Step 3: Determination of the Lifetime Health Advisory

$$\text{Lifetime HA} = (0.2 \text{ mg/L}) (20\%) = 0.04 \text{ mg/L } (40 \text{ }\mu\text{g/L})$$

Note: The NTP (1988) is conducting carcinogenicity studies for 1,2,3-TCP in animals. The Lifetime HA for 1,2,3-TCP will be reevaluated following a review of the results on the NTP bioassay in animals when made available by the NTP.

E. Evaluation of Carcinogenic Potential

- The carcinogenic potential of 1,2,3-TCP has not been reported, but this chemical is being tested for carcinogenicity by the NTP (1988). IARC has not evaluated the carcinogenic potential of 1,2,3-TCP. The evaluation for carcinogenic potential is being deferred until the completion of the NTP studies.

V. Other Criteria, Guidance and Standards

- ACGIH (1980, 1985) recommended a Threshold Limit Value (TLV) of 50 ppm (300 mg/m³) to prevent hepatotoxicity caused by 1,2,3-TCP, which is typical of many chlorinated hydrocarbons. ACGIH (1980, 1985) recommended a Short-term Exposure Level (STEL) of 75 ppm (450 mg/m³) to prevent eye and mucosal irritation. ACGIH (1985) proposed changing the TLV to 10 ppm (60 mg/m³), but no reason was given.

- The OSHA Permissible Exposure Limit (PEL) for 1,2,3-TCP is 50 ppm (300 mg/m³) (CFR, 1985).

VI. Analytical Methods

- Analysis of 1,2,3-TCP is by a purge-and-trap gas chromatographic procedure used for the determination of volatile organohalides in drinking water (U.S. EPA, 1985a). This method calls for the bubbling of an inert gas through the sample and trapping volatile compounds on an adsorbent material. The adsorbent material is heated to drive off the compounds onto a gas chromatographic column. The gas chromatograph is temperature programmed to separate the method analytes, which are then detected by a halogen specific detector. Confirmatory analysis is by mass spectrometry (U.S. EPA, 1985b). The detection limit has not been determined for either method.

VII. Treatment Technologies

- Leighton and Calo (1981) reported experimental measurements of the distribution coefficients for 21 chlorinated hydrocarbons, including 1,2,3-trichloropropane, in a dilute air-water system. (The distribution coefficient is the ratio of the volume of the compound in air to the volume of the compound in water after purging). They determined the distribution coefficient for 1,2,3-trichloropropane to be approximately 20 at 25°C.

- U.S. EPA (1986a) estimated the feasibility of removing 1,2,3-trichloropropane from water by packed column aeration, employing the engineering design procedure and cost model presented at the 1983 National ASCE Conference on Environmental Engineering. Based on chemical and physical properties and assumed operating conditions, a 90% removal efficiency of 1,2,3-trichloropropane was reported for a column with a diameter of 6.7 feet and packed with 16 feet of 1-inch plastic saddles. The air-to-water ratio required to achieve this degree of removal effectiveness is 40.

- No data were presented for the removal of 1,2,3-trichloropropane from drinking water by activated carbon adsorption. However, evaluation of physical/chemical properties indicates that it may be amenable to removal by activated carbon adsorption due to its low solubility.

VIII. References

ACGIH. 1980. American Conference of Governmental Industrial Hygienists. Documentation of the threshold limit values, 4th ed. Cincinnati, OH: ACGIH. pp. 410–411.

ACGIH. 1985. American Conference of Governmental Industrial Hygienists. TLVs. Threshold limit values and biological exposure indices for 1985–86. Cincinnati, OH: ACGIH. pp. 32, 37.

Belyaeva, N.N., T.I. Bonashevskaya, T.L. Marshak and V.Y.A. Brodskii. 1977. Investigation of the effect of certain chlorinated hydrocarbons on the composition of the hepatocyte population of the rat liver. Bull. Exp. Biol. Med. (USSR). 83:396–400. (Cited in U.S. EPA, 1983.)

CFR. 1985. Code of Federal Regulations. OSHA Occupational Standards. Permissible Exposure Limits. 29 CFR 1910.1000.

Dewalle, F.B. and E.S.K. Chian. 1978. Presence of trace organics in the Delaware River and their discharge by municipal and industrial source. Proc. Ind. Waste Conf. 32:908–919.

Dilling, W.L. 1977. Interphase transfer processes. II. Evaporation rates of chloromethanes, ethanes, ethylenes, propanes, and propylenes from dilute aqueous solutions. Comparison with theoretical predictions. Environ. Sci. Technol. 11:405–409.

Hardin, B.D., G.P. Bond, M.R. Sikov, F.D. Andrew, R.P. Beliles and R.W. Niemeier. 1981. Testing of selected workplace chemicals for teratogenic potential. Scand. J. Work, Environ. Health, 7(Suppl. 4). pp. 66–75.

Johannsen, F.R., G.J. Leumskas, G.M. Rusch, J.B. Terrill and R.E. Schroeder. 1988. Evaluation of the subchronic and reproductive effects of a series of chlorinated propanes in the rat. I. Toxicity of 1,2,3-trichloropropane. J. Tox. Env. Health. 25:299–315.

Keith, L.W., A.W. Garrison, F.R. Allen et al. 1976. Identification of organic compounds in drinking water from thirteen U.S. cities. In: Keith, L.H., ed. Identification and Analysis of Organic Pollutants in Water. Ann Arbor, MI: Ann Arbor Science. pp. 329–373.

Koneman, H. 1981. Qualitative structure-activity relationships in fish (*Poecilia reticulata*) toxicity studies. I. Relationship for 50 industrial pollutants. Toxicol. 19(3):209–225.

Kool, H.J., C.F. Van Kreijl and B.C.J. Zoeteman. 1982. Toxicology assessment of organic compounds in drinking water. Crit. Rev. Environ. Control. 12:307–357.

Leighton, D.T., Jr. and J.M. Calo. 1981. Distribution coefficients of chlorinated hydrocarbons in dilute air-water systems for ground water contamination applications. J. Chem. Eng. Data. 26:382–385.

Lewis, T.R. 1979. Personal communication to TLV Committee, Cincinnati, OH. (Cited in ACGIH, 1980.)

Matsui, S., T. Murakami, T. Sasaki, Y. Hirose and Y. Iguma. 1975. Activated sludge degradability of organic substances in the waste water of the Kashima petroleum and petrochemical industrial complex in Japan. Prog. Water Technol. 7:645–649.

McOmie, W.A. and T.R. Barnes. 1949. Acute and subacute toxicity of 1,2,3-trichloropropane in mice and rabbits. Fed. Proc. 8:319.

NIOSH. 1986. National Institute of Occupational Safety and Health. RTECS (Registry of Toxic Effects of Chemical Substances). February 1986: Online.

NTP. 1983a. National Toxicology Program. Final report: 120-day toxicity gavage study of 1,2,3-trichloropropane in Fischer 344 rats. Vienna, VA: Hazelton Laboratories America, Inc. June 16.

NTP. 1983b. National Toxicology Program. Final report: 120-day gavage toxicity study in B6C3F$_1$ mice. 1,2,3-Trichloropropane. Vienna, VA: Hazelton Laboratories America, Inc. April 29.

NTP. 1988. National Toxicology Program. Toxicology Research and Testing Program. Management Status Report, 10/12/88. Research Triangle Park, NC: NTP.

RTECS. 1978. Registry of toxic effects of chemical substances. Cincinnati, OH: U.S. Dept. of Health, Education and Welfare. Centers for Disease Control. National Inst. Occ. Safety and Health.

Saito-Suzuki, R., S. Teramoto and Y. Shirasu. 1982. Dominant lethal studies in rats with 1,2-dibromo-3-chloropropane and its structurally related compounds. Mutat. Res. 101(4):321–327.

Shcherban, N.G. and N.N. Piten'ko. 1976. Effect of trichloropropane and pentachloropropane on the body of warm-blooded animals. Tr. Khar'k. Med. Inst., Volume Date 1975. 124:27–29. [CA 89(23):191916y]

Sidorenko, G.I., V.R. Tsulaya, T.I. Bonashevskaya and V.M. Shaipak. 1979. Study of the combined action of a group of chlorine derivatives of hydrocarbons entering the organism by inhalation. Environ. Health Perspect. 30:13–18.

Silverman, L., H.F. Schutte and M.W. First. 1946. Further studies on sensory response to certain industrial solvent vapors. J. Ind. Hyg. Toxicol. 28:262–266.

Smythe, H.F., Jr., C.P. Carpenter, C.S. Weil, U.C. Pozzani and J.A. Striegel. 1962. Range-finding toxicity data: List VI. J. Am. Ind. Hyg. Assoc. 23:95–107.

Stolzenberg, S.J. and C.H. Hine. 1980. Mutagenicity of 2- and 3-carbon halogenated compounds in the *Salmonella*/mammalian microsome test. Environ. Mutagen. 2(1):59–66.

Tarasova, K.I. 1975. Morphofunctional changes in mast cells caused by 1,2,3-trichloropropane and tetrachloroethylene. Gig. Sanit. pp. 106–109. [CA 84(11):69971u]

U.S. EPA. 1976. Frequency of organic compounds identified in water. Prepared by Environmental Research Laboratory. Athens, GA: ORD. p. 205.

U.S. EPA. 1983. Hazard profile for 1,2,3-trichloropropane. Prepared by the Office of Health and Environmental Assessment, Environmental Criteria and Assessment Office, Cincinnati, OH, for the Office of Solid Waste, Washington, DC.

U.S. EPA. 1985a. Method 502.1–Volatile halogenated organic compounds in water by purge and trap gas chromatography. Cincinnati, OH: Environmental Monitoring and Support Laboratory. June 1985 (Revised November 1985).

U.S. EPA. 1985b. Method 524.1–Volatile organic compounds in water by purge and trap gas chromatography/mass spectrometry. Cincinnati, OH:

Environmental Monitoring and Support Laboratory. June 1985 (Revised November 1985).

U.S. EPA. 1986a. Economic evaluation of 1,2,3-trichloropropane removal from water by packed column air stripping. Prepared by Office of Water for Health Advisory Treatment Summaries.

U.S. EPA. 1986b. Guidelines for carcinogen risk assessment. *Fed. Reg.* 51(185):3392–3403. Sept. 24.

Verschueren, K. 1977. Handbook of Environmental Data on Organic Chemicals. New York, NY: Van Nostrand Reinhold Co. p. 613.

Volp, R.F., I.G. Sipes, C. Falcoz, D.E. Carter and J.F. Gross. 1984. Disposition of 1,2,3-trichloropropane in the Fischer 344 rat: conventional and physiological pharmacokinetics. Toxicol. Appl. Pharmacol. 75(1):8–17.

Wakeham, S.T., J.T. Goodwin and A.C. Davis. 1983. Distribution and fate of volatile organic compounds in Narragansett Bay, RI. Can. J. Fish. Aquatic Sci. 40(2):304–321.

Weast, R.C. 1972. Handbook of Chemistry and Physics. Cleveland, OH: The Chemical Rubber Co.

Wright, W.H. and J.M. Schaffer. 1932. Critical anthelmintic tests of chlorinated alkyl hydrocarbons and a correlation between the anthelmintic efficacy. Chemical structure and physical properties. Am. J. Hyg. 16:325–428.

Bromomethane

I. General Information and Properties

A. CAS No. 74–83–9

B. Structural Formula

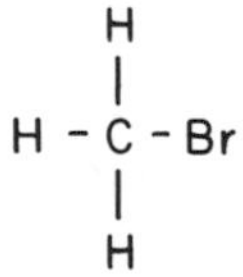

```
        H
        |
  H  -  C  - Br
        |
        H
```

Bromomethane

C. Synonyms

- Methyl bromide, Brom-O-Gas, Grom-O-Gaz, Brom-O-Sol, Celfume, Kayafume, Meth-O-Gas, Terr-O-Cide II, Terr-O-Gas (Meister, 1988).

D. Uses

- Bromomethane is used primarily as a fumigant in soil to control fungi, nematodes and weeds (65% of use) and in the space fumigation of food commodities and storage facilities to control insects and rodents (15% of use) (Stenger, 1978; Worthing and Walker, 1983; CMR, 1985). Bromomethane is also used in the chemical industry as a methylating agent in the preparation of aniline dyes and other compounds, as an extraction solvent for vegetable oils (e.g., nuts, seeds and flowers) and for degreasing wool (Hawley, 1981).

E. **Properties** (Stenger, 1978; Verschueren, 1983; Hansch and Leo, 1985; Ruth, 1986)

Chemical Formula	CH_3Br
Molecular Weight	94.94
Physical State (at 25°C)	Gas
Boiling Point	3.6°C
Melting Point	-93.7°C
Density (0°C)	1.73 g/mL
Vapor Pressure (20°C)	1,420 mm Hg
Specific Gravity (at 0°C)	1.73
Water Solubility (20°C)	17,500 mg/L
Log Octanol/Water Partition Coefficient	1.19
Taste Threshold	—
Odor Threshold (water)	—
Odor Threshold (air)	80 mg/m^3
Conversion Factor	1 ppm = 4 mg/m^3
	1 mg/m^3 = 0.25 ppm

F. **Occurrence**

- Bromomethane has been detected in drinking water (Kool et al., 1982; U.S. EPA, 1975; Shackelford and Keith, 1976), ground water (Greenberg et al., 1982) and seawater (Singh et al., 1977, 1983; Lovelock, 1975). Bromomethane was not detected in storm water runoff monitored in 15 U.S. cities (Cole et al., 1984).

- Gould et al. (1983) tentatively identified bromomethane in chlorinated landfill leachate and suggested that it is formed by the chlorination process, whereby natural bromine is reduced by hypochlorous acid to hypobromous acid, which is capable of forming bromomethane. Water chlorination treatment, therefore, may be a source of bromomethane found in finished drinking water. The source of bromomethane in seawater appears to be natural formation, although the mechanism for formation is unclear (Singh et al., 1977, 1983).

- Staples et al. (1985) reported that 1.4% of the bromomethane analyses from the U.S. EPA STORET data base for industrial effluents and ambient water were positive, with the median concentration of the positive analyses equal to < 10 μg/L.

- Atmospheric monitoring studies have detected bromomethane in urban air, rural air, oceanic air, arctic air, stratospheric air and air near manufacturing plants (U.S. EPA, 1986b). Sources of emission include its use as a soil or space fumigant (U.S. EPA, 1985a) and release from automobile exhaust (Harsch and Rasmussen, 1977). Formation and release from seawaters may be the dominant source of atmospheric bromomethane (Singh et al., 1983). The major source of bromomethane in urban areas may be auto exhaust.

- Extensive monitoring of foods by Federal programs has resulted in infrequent detection of bromomethane (Duggan et al., 1983). Bromomethane disappears from fumigated foods within a few days, leaving only inorganic bromide residues (Stenger, 1978).

G. Environmental Fate

- Volatilization half-lives at 25°C of 1, 3.9 and 5 days were calculated for river, lake and pond water, respectively (Mabey et al., 1981).

- The hydrolytic half-life of bromomethane in 20° to 25°C water has been calculated from experimental data to be 20 to 38 days (Mabey and Mill, 1978; Castro and Belser, 1981; Ehrenberg et al., 1974), indicating that hydrolysis is likely to be a significant removal process (U.S. EPA, 1986b).

- Photolysis, oxidation, biodegradation, adsorption and bioaccumulation of bromomethane in aqueous environments are not expected to be significant (U.S. EPA, 1986b).

II. Pharmacokinetics

A. Absorption

- Seventy-two hours after male Fischer 344 rats ingested 250 μmoles ^{14}C-bromomethane/kg (23.7 mg/kg), in corn oil, 32 and 43% of the administered radioactivity was recovered from expired air and urine, respectively; < 3% was found in the feces and 14% was found in the carcass (Medinsky et al., 1984), indicating extensive absorption of ^{14}C associated with ^{14}C-bromomethane after oral administration. Most of the urinary and pulmonary excretion of radioactivity occurred within 20 hours.

- Of inhaled ^{14}C-bromomethane, 50% was absorbed by male Fischer 344 rats exposed to 50 or 300 nmoles/L (4.7 or 28 mg/m^3) for 6 hours. The amount absorbed was determined by measuring carcass retention of radioactivity at the end of exposure (Medinsky et al., 1985). At higher exposure levels (541 and 987 mg/m^3), a smaller percentage (38 and 27%, respectively) of the inhaled radioactivity was absorbed. In light of a conclusion by Kornbrust and Bus (1983) that glutathione conjugation may be a major metabolic pathway for methyl halides, Medinsky et al. (1985) suggested that the availability of glutathione may be the rate-limiting factor on the pulmonary absorption of bromomethane at higher exposure concentrations. The decreased rate of absorption with increased exposure levels indicates that the uptake of inhaled bromomethane could be saturated.

B. Distribution

• Seventy-two hours after oral or intraperitoneal administration of 250 μmoles ^{14}C-bromomethane/kg (23.7 mg/kg) to male Fischer 344 rats, the highest concentration of radioactivity was found in the liver, followed by the kidney, testes, lung, heart, stomach and spleen (Medinsky et al., 1985). There was little difference in the tissue concentration of radioactivity due to the route of administration except in the stomach, which had a greater amount of residual radioactivity after oral administration.

• Sixty hours after inhalation exposure of male Fischer 344 rats to 337 nmoles/L (32 mg/m^3) of ^{14}C-bromomethane for 6 hours, the highest concentration of radioactivity was found in the liver, followed by the kidneys, stomach, thymus, lungs and adrenals (Bond et al., 1985). The longest half-life of radioactivity occurred in the liver (33.0 ± 4.9 hours), followed by the blood (7.7 ± 0.4 hours) and testes (6.2 ± 0.2 hours); whereas the shortest half-life occurred in the adrenals (1.6 ± 0.1 hours).

C. Metabolism

• Based on its chemical similarity to other methyl halides (i.e., methyl chloride), Medinsky et al. (1985) concluded that bromomethane reacts with sulfhydryl groups, particularly with the sulfhydryl groups on hepatic glutathione, and is then excreted in the bile as the conjugate S-methylglutathione, which is subsequently reabsorbed in the small intestines. After reabsorption, methylglutathione may be metabolized to a number of intermediates, such as N-acetyl-S-methylcysteine or methyl-thioacetic sulfoxide, which is excreted in the urine, or to S-methylcysteine, which is further metabolized to formate with subsequent production of CO_2 (Kornbrust and Bus, 1983; Medinsky et al., 1985).

D. Excretion

• During the 72 hours after oral administration of 250 μmoles ^{14}C-bromomethane/kg (23.7 mg/kg) to male Fischer 344 rats, 43% of the administered radioactivity was recovered from the urine, 36% from the expired air (32% as CO_2 and 4% as bromomethane) and $< 3\%$ from the feces (Medinsky et al., 1984).

- The importance of biliary excretion in rats was demonstrated by Medinsky et al. (1984), who administered orally 250 μmoles ^{14}C-bromomethane/kg (23.7 mg/kg) to bile duct cannulated rats and collected 46% of the administered dose of radioactivity in the bile within 25 hours. The importance of enterohepatic recirculation was shown by comparing the 24-hour excretion of these bile duct cannulated rats with that of similarly treated intact rats. Exhaled CO_2, urine and feces accounted for 29, 38 and 0.5%, respectively, of the administered dose of radioactivity in intact rats. In bile duct cannulated rats, exhaled CO_2 and urine accounted for 12 and 7% of the administered dose, respectively, and the amount of administered dose recovered in the feces was below the detection limit.

- During the 65 hours after inhalation exposure of male rats to 337 nmoles/L (32 mg/m³) of ^{14}C-bromomethane for 6 hours, 46.7% of the administered radioactivity was recovered as CO_2 and 1.0% as bromomethane from the expired air, 22.7% from the urine and 2.0% from the feces (Bond et al., 1985). Urinary and fecal excretion of radioactivity appeared to follow first-order kinetics with half-lives of 9.6 and 16.1 hours, respectively. Elimination of radioactivity through the lungs appeared to follow second-order kinetics, with half-lives of 3.9 and 11.4 hours for the α and β phases, respectively. These conclusions were essentially confirmed by the work of Medinsky et al. (1985).

III. Health Effects

A. Humans

1. Short-term Exposure

- Bromomethane is a highly toxic chemical. Because it is a gas at 4°C and atmospheric pressure, most human exposure has occurred through inhalation. As summarized by U.S. EPA (1986b), fatal poisoning has occurred in humans exposed to 300 to 400 ppm of bromomethane for short periods of time (Hine, 1969). The ACGIH (1980) reported that workers exposed to 35 ppm for 2 weeks developed mild systemic poisoning accompanied by nausea, vomiting, headache and skin lesions. The onset of symptoms may be delayed several hours after exposure and recovery is often slow. The primary targets of bromomethane exposure in humans exposed through inhalation are the respiratory, nervous and gastrointestinal systems. The dose-response curve is quite steep for bromomethane. Degenerative changes may also be seen in the kidneys, liver and skin after acute human exposure to bromomethane. Pulmonary edema may result in death within hours after bromomethane exposure, whereas circulatory failure is the precipitating cause of death if a victim survives longer (Torkelson et al., 1966; Longley and Simpson, 1970; Araki et al., 1971; Greenberg, 1971; Ushio and Osozuka, 1977; Zatuchni and Hong, 1981).

2. Dermal/Ocular Effects

- Bromomethane can be absorbed through the skin of humans to produce adverse neurological effects similar to those produced after inhalation exposure (Longley and Jones, 1965). Although it is poorly absorbed through the skin, burning and irritation can occur (Torkelson et al., 1966).

- Ophthalmological disturbances were also noted in workers engaged in bromomethane preparation for 5 months to 10 years (Ishitsu et al., 1975). The exposure level and a correlation between the observed effect and duration of exposure were not reported.

3. Long-term Exposure

- The abstract of a Japanese study (Ishitsu et al., 1975) reported neurological disturbances in workers engaged in bromomethane preparation for 5 months to 10 years. Actual exposure levels and a correlation between observed effects and duration of exposure were not reported. The U.S. EPA (1985a) reported that the available occupational exposure data are not adequate for quantitative analysis.

- The incidence of death due to testicular cancer was significantly higher (2/665 observed; 0.11/665 expected) in workers exposed to organic bromides (Wong et al., 1984). The two men who were exposed to bromomethane may also have been exposed to other chemicals. Exposures were not quantitated.

B. Animals

1. Short-term Exposure

- Dudley and Neal (1942) found the minimum single lethal oral dose of bromomethane for rabbits to be 60 to 65 mg/kg body weight; Miller and Haggard (1943) observed that rats died 5 to 7 hours after an oral dose of 100 mg/kg body weight; and Danse et al. (1984) reported an oral LD_{50}

for rats to be 214 mg/kg body weight. The differences in the results might be due to a difference in strain of rats, and solvent used.

- Alexeeff et al. (1985) reported that there were no adverse effects in male Swiss-Webster mice after inhalation of ≤ 1.72 mg/L ($\leq 1,720$ mg/m^3) bromomethane for 1 hour, but that significant decreases in body, lung and liver weight, kidney and liver lesions, abnormal clinical signs and increased mortality were observed in mice exposed to higher dose levels. One hour LC_{50} for mouse via inhalation is 4.68 mg/L (1,200 ppm). The length of the observation period after treatment was not reported.

- Norepinephrine levels were significantly depressed in the brain in groups of 5 to 6 male albino rats exposed to 100 to 120 ppm (388 to 466 mg/m^3), but not ≤ 60 ppm (233 mg/m^3), of bromomethane for 24 hours (Honma et al., 1982). The decrease was still significant 24 hours after the end of treatment.

- Norepinephrine levels were significantly decreased in the brains of rats exposed to 5 or 10 ppm (10 or 39 mg/m^3), but not 1 ppm (4 mg/m^3), of bromomethane continuously for 3 weeks by inhalation (Honma et al., 1982). Serotonin levels were also significantly depressed at 10 ppm (39 mg/m^3).

2. Dermal/Ocular Effects

- No data were available on the dermal or ocular effects of bromomethane.

3. Long-term Exposure

- Treatment of groups of 10 male and 10 female Wistar rats by gavage 5 days/week for 13 weeks with bromomethane at 0, 0.4, 2, 10 or 50 mg/kg resulted in severe hyperplasia (the author considered this as squamous cell carcinoma but NTP disagreed. See p. 9) of the stratified squamous epithelium in the forestomach at a dose of 50 mg/kg/day and slight epithelial hyperplasia in the forestomach at a dose of 10 mg/kg/day (Danse et al., 1984). At the 50 mg/kg/day dose level, decreased food consumption, body weight gain and anemia were observed in the male rats. Slight pulmonary atelectasis was observed at the two higher dose levels, in both male and female rats; however, the investigators stated that the possible inhalation of bromomethane-containing oil during the gastric intubation procedure might have been responsible for this effect. Decreased hemosiderosis and increased hematopoiesis in the spleen of males of the highest dose were noticed. No treatment-related liver or neurotoxic effects were observed. Renal histopathology was not evaluated. No adverse effects were observed at 0.4 or 2 mg/kg.

- Groups of rats, guinea pigs, rabbits and monkeys were exposed by inhalation to various levels of bromomethane for 7.5 to 8 hours/day, 5 days/week for up to 6 months (Irish et al., 1940). Pulmonary lesions and death occurred in guinea pigs exposed to 0.85 mg/L (220 ppm). Convulsions, weight loss and pathological changes in the lungs of rats were observed at 0.42 mg/L (100 ppm). At a concentration of 0.13 mg/L (33 ppm), paralysis and pulmonary disease were noted in rabbits. Exposure to levels of 0.25 mg/L (66 ppm) resulted in paralysis and convulsions in monkeys. No adverse effects were noted in rabbits at 0.065 mg/L (17 ppm), in monkeys at 0.13 mg/L (33 ppm), in rats at μ0.25 mg/L (μ66 ppm) or in guinea pigs at 0.42 mg/L (100 ppm).

4. Reproductive Effects

- No data were located in the available literature regarding the reproductive effects of oral exposure to bromomethane. However, by the inhalation route, there is a two-generation reproductive study by the American Biogenics Corporation identifying a NOAEL of 3 ppm (U.S. EPA, 1988).

5. Developmental Effects

- No adverse developmental effects were observed in the fetuses of Wistar rats exposed to 20 ppm (78 mg/m^3) or 70 ppm (272 mg/m^3) of bromomethane in the air for 7 hours/day on days 1 through 19 of gestation (Hardin et al., 1981; Sikov et al., 1980). Exposure to 20 ppm (78 mg/m^3) or 70 ppm (272 mg/m^3) for 7 hours/day, 5 days/week for 3 weeks prior to mating, and through gestation, did not result in developmental toxicity. Maternally toxic effects were not observed.

- Bromomethane was highly toxic to pregnant New Zealand White rabbits exposed to 70 ppm (272 mg/m^3) in the air for 7 hours/day, 5 days/week on days 1 through 15 of gestation; 24/25 rabbits died by day 30 of gestation (Hardin et al., 1981; Sikov et al., 1980). No adverse developmental effects were observed in the one remaining litter or in the group of rabbits exposed to 20 ppm (78 mg/m^3) of bromomethane for 7 hours/day, 5 days/week on days 1 through 30 of gestation.

6. Mutagenicity

- Bromomethane was mutagenic to Salmonella typhimurium (strains TA1535 and TA100, but not strains TA1537 and TA98), when tested at vapor concentrations μ50 g/m^3, with or without metabolic activation (Kramers et al., 1985; Moriya et al., 1983; Simmon et al., 1977; Simmon and Tardiff, 1978; Voogd et al., 1982). In a gene reversion assay with

Escherichia coli, bromomethane was not mutagenic when tested at vapor concentrations $\mu50$ g/m^3 (Moriya et al., 1983).

- Bromomethane was mutagenic in recessive lethal tests in *Drosophila melanogaster* at concentrations of 200 and 375 mg/m^3 for 90 and 30 hours, respectively (Voogd et al., 1982), 20 to 70 ppm for 5 hours (McGregor, 1981) and 70 to 750 mg/m^3 for 6 hours/day, 5 days/week for 1 to 3 weeks (Kramers et al., 1985).

- Bromomethane was mutagenic in the mouse lymphoma cell assay at concentrations of ≥ 0.3 mg/L (Voogd et al., 1982). In unscheduled DNA synthesis assays in rat liver cells, it was not mutagenic at concentrations of 10 to 30 mg/L (Voogd et al., 1982) or 4 mg/mL (Kramers et al., 1985). Bromomethane did not elicit unscheduled DNA synthesis in human fibroblasts when tested as a vapor at concentrations $\leq 70\%$ for 3 hours, with or without metabolic activation (McGregor, 1981).

- Bromomethane was not mutagenic in *in vivo* assays in rats or mice exposed by inhalation to 0, 20 or 70 ppm (0, 78 or 272 mg/m^3) of bromomethane for 7 hours/day for 5 days (McGregor, 1981). No sperm abnormalities were observed in B6C3F$_1$ mice or Sprague-Dawley rats; no chromosomal abnormalities were observed in bone marrow cells of male or female Sprague-Dawley rats.

7. Carcinogenicity

- Although an increased incidence of carcinoma of the forestomach occurred in 6/10 females and 7/10 males treated by gavage with bromomethane dissolved in arachis oil at a concentration of 50 mg/kg/day, 5 days/week for 13 weeks (Danse et al., 1984), reevaluation of the histological slides by NTP scientists led to the conclusion that the reported lesions in the forestomachs were hyperplastic/inflammatory lesions and not carcinomas (NTP, 1984). Since the study was inconclusive, bromomethane is classified in Group D.

IV. Quantification of Toxicological Effects

A. One-day Health Advisory

Data are inadequate for derivation of a One-day HA for bromomethane. Single dose or acute oral studies were not located in the available literature. Single exposure inhalation studies have been performed in mice (Alexeeff et al., 1985) and rats (Honma et al., 1982). Signs of central nervous system damage appear to be a common manifestation to toxicity due to inhalation

(ACGIH, 1980; Honma et al., 1982) but not oral (Danse et al., 1984) exposure. The major target organ system for oral exposure appears to be the gastrointestinal tract. There also appears to be marked differences in the pharmacokinetics of orally administered bromomethane and inhaled bromomethane (Medinsky et al., 1984, 1985; Bond et al., 1985). It seems inappropriate, therefore, to derive an HA for bromomethane based on inhalation exposure. The Longer-term HA of 0.1 mg/L (calculated below) for a 10-kg child is used for the One-day HA for bromomethane.

B. Ten-day Health Advisory

Data are inadequate for derivation of a Ten-day HA for bromomethane. The Longer-term HA of 0.1 mg/L for a 10-kg child, calculated below, is used for the Ten-day HA for bromomethane.

C. Longer-term Health Advisory

The study by Danse et al. (1984) has been selected to serve as the basis for the Longer-term HA because it provides the only well-quantified oral data on the toxicity of bromomethane. In the Danse et al. (1984) study, treatment of groups of 10 male and 10 female rats 5 days/week for 13 weeks by gavage with 0.4, 2, 10 or 50 mg bromomethane/kg resulted in slight hyperplasia of the stratified squamous epithelium in the forestomachs of rats treated with 10 mg/kg, and severe hyperplasia and inflammation in the forestomach of rats treated with 50 mg/kg. The highest NOAEL in the Danse et al. (1984) study is 2 mg/kg/day, corresponding to a Time-weighted Average of 1.4 mg/kg/day (2 mg/kg/day × 5/7). This NOAEL serves as the basis for the Longer-term HA.

The Longer-term HA for a 10-kg child is calculated as follows:

$$\text{Longer-term HA} = \frac{(1.4 \text{ mg/kg/day}) (10 \text{ kg})}{(100) (1 \text{ L/day})} = 0.14 \text{ mg/L (rounded to } 100 \text{ } \mu\text{g/L)}$$

The Longer-term HA for a 70-kg adult is calculated as follows:

$$\text{Longer-term HA} = \frac{(1.4 \text{ mg/kg/day}) (70 \text{ kg})}{(100) (2 \text{ L/day})} = 0.49 \text{ mg/L (rounded to } 500 \text{ } \mu\text{g/L)}$$

D. Lifetime Health Advisory

The study by Danse et al. (1984) has been selected to serve as the basis for the Lifetime HA because, despite its short duration (13 weeks), it provides the only well-quantified oral data on the toxicity of bromomethane.

Step 1: Determination of the Reference Dose (RfD)

$$RfD = \frac{(1.4 \text{ mg/kg/day})}{(100) \ (10)} = 0.0014 \text{ mg/kg/day}$$

where:

10 = additional uncertainty factor to compensate for less-than-lifetime exposure.

Step 2: Determination of the Drinking Water Equivalent Level (DWEL)

$$DWEL = \frac{(0.0014 \text{ mg/kg/day}) \ (70 \text{ kg})}{(2 \text{ L/day})} = 0.049 \text{ mg/L} \ (50 \ \mu g/L)$$

Step 3: Determination of the Lifetime Health Advisory

Lifetime HA = (0.049 mg/L) (20%) = 0.01 mg/L (10 μg/L)

E. Evaluation of Carcinogenic Potential

- There are currently in progress two chronic studies, one sponsored by the NTP and one at the Dutch National Institute for Public Health (NTP, 1987; U.S. EPA, 1985a) that may provide a basis for evaluating the carcinogenicity of bromomethane.

- IARC (1987) has categorized bromomethane as Group 3: not classifiable.

- Applying the criteria described in the U.S. EPA's guidelines for carcinogen risk assessment (U.S. EPA, 1986a), bromomethane may be classified in Group D: not classified. This category is for agents with inadequate animal evidence of carcinogenicity.

V. Other Criteria, Guidance and Standards

- A Threshold Limit Value (TLV) of 5 ppm (20 mg/m^3), with a Short-term Exposure Level (STEL) of 15 ppm (60 mg/m^3), was recommended by the ACGIH (1980) to protect against the adverse neurotoxic and pulmonary effects observed after acute exposure. Bromomethane is designated with the skin notation indicating the potential contribution of dermal absorption.

- The OSHA standard is 20 ppm (80 mg/m^3) for an 8-hour Time-weighted Average (TWA) exposure limit (CFR, 1985).

- Tolerances for bromomethane on raw agricultural commodities range from 5 ppm on pears, apples and quinces to 300 ppm on asparagus and lettuce (U.S. EPA, 1982, 1983).

VI. Analytical Methods

- Analysis of bromomethane is by a purge-and-trap gas chromatographic procedure used for the determination of volatile organohalides in drinking water (U.S. EPA, 1985a). This method calls for the bubbling of an inert gas through the sample and trapping volatile compounds on an adsorbent material. The adsorbent material is heated to drive off the compounds onto a gas chromatographic column. The gas chromatograph is temperature-programmed to separate the method analytes, which are then detected by a halogen specific detector. Confirmatory analysis is by mass spectrometry (U.S. EPA, 1985b). The detection limit has not been determined for either method.

VII. Treatment Technologies

- A Henry's Law constant of 788 atm was calculated based on the reported vapor pressure (Perry and Chilton, 1973) and water solubility of 13.4 g/L (Worthing and Walker, 1983) at 25°C. Based on the Henry's Law constant and by comparison with compounds which are readily removed from water by packed column aeration (i.e., carbon dioxide, Henry's Law constant 1.5×10^3 atm), it can be concluded that bromomethane will be removed from water by aeration.

- The efficiency of powdered activated carbon (PAC) in removing volatile organics, including bromomethane, was investigated at DuPont's Chambers Works Wastewater Treatment Plant (Dunn and Hutton, 1983). Nuchar SA-15 activated carbon was used at a dosage of 117 mg/L for the treatment of 36 million gallons per day (MGD) of wastewater containing 169 mg/L soluble Total Organic Carbon (TOC). PAC was fed upstream of the aeration chamber designed for 8 hours aeration time. The results show 95% removal efficiency for bromomethane. The concentration of bromomethane in the untreated wastewater is not specified. The data presented in this study, however, are not conclusive. Bromomethane is a highly volatile compound. Therefore, the excellent removal efficiency reported in this study might be the result of the aeration as well as adsorption on activated carbon. Aeration is the probable mechanism of removal.

- Due to the high vapor pressure and solubility of bromomethane, GAC may be less effective than air stripping in removing bromomethane from drinking water.

VIII. References

ACGIH. 1980. American Conference of Governmental Industrial Hygienists. Documentation of the threshold limit values for chemical substances in the workroom air, 4th ed. Cincinnati, OH. pp. 265–266.

Alexeeff, G.V., W.W. Kilgore, P. Munoz and D. Watt. 1985. Determination of acute toxic effects in mice following exposure to methyl bromide. J. Toxicol. Environ. Health. 15:109–123.

Araki, S., K. Ushio, K. Suwa, A. Abe and K. Uehara. 1971. Methyl bromide poisoning: a report based on fourteen cases. Sangyo Igaku. 13(6):507–513. Data base abstr. from Health Aspects Pest. Abstr. Bull. 72:0327.

Bond, J.A., J.S. Dutcher, M.A. Medinsky, R.F. Henderson and L.S. Birnbaum. 1985. Disposition of [^{14}C]-methyl bromide in rats after inhalation. Toxicol. Appl. Pharmacol. 78:259–267.

Castro, C.F. and N.O. Belser. 1981. Photohydrolysis of methyl bromide and chloropicrin. J. Agric. Food Chem. 29:1005–1008.

CFR. 1985. Code of Federal Regulations. OSHA occupational standards: permissible exposure limits. 29 CFR 1910.1000.

CMR. 1985. Chemical Marketing Reporter. Chemical profile—methyl bromide. February 18.

Cole, R.H., R.E. Frederick, R.P. Healy and R.G. Rolan. 1984. Preliminary findings of the priority pollutant monitoring project of the Nationwide Urban Runoff Program. J. Water Pollut. Control Fed. 56:898–908.

Danse, L.H.J.C., F.L. van Velsen and C.A. van der Heijden. 1984. Methylbromide: carcinogenic effects in the rat forestomach. Toxicol. Appl. Pharmacol. 72:262–271.

Dean, J.A. 1979. Lange's Handbook of Chemistry. New York, NY: McGraw Hill Book Co.

Dudley, H.C. and P.A. Neal. 1942. Methyl bromide as a fumigant for feeds. Food Res. 7:412–429.

Duggan, R.E., P.E. Corneliussen, M.B. Duggan, B.M. McMahon and R.J. Martin. 1983. Pesticide residue levels in foods in the United States from July 1, 1969 to June 30, 1976. Washington, DC: U.S. FDA, DCT. pp. 14–15.

Dunn, G.F. and D.G. Hutton. 1983. The combined powdered activated carbon-activated sludge (PACT) process for toxics control. Toxic Materials Methods for Control. Water Resources Symposium No. 10. Amstrong,

N.E. and A. Kudo, eds. Center for Research and Water Resources, University of Texas.

Ehrenberg, L., S. Osterman-Golkar, D. Singh and U. Lundqvist. 1974. On the reaction kinetics and mutagenic activity of methylating and β-halogenoethylating gasoline additives. Radiat. Bot. 15:185–194.

Gould, J.P., R.E. Ramsey, M. Giabbai and F.G. Pohland. 1983. Formation of volatile haloorganic compounds in the chlorination of municipal landfill leachates. In: R.L. Jolley et al., ed. Water Chlorination: Environmental Impact and Health Effects, Vol. 4. Seven Oaks, England: Ann Arbor Science, The Butterworth Group, pp. 525–539.

Greenberg, J.O. 1971. The neurological effects of methyl bromide poisoning. Ind. Med. 49(4):27–29.

Greenberg, M., R. Anderson, J. Keene, A. Kennedy, G.W. Page and S. Schowgurow. 1982. Empirical test of the association between gross contamination of wells with toxic substances and surrounding land use. Environ. Sci. Technol. 16:14–19.

Hansch, C. and A.S. Leo. 1985. Medchem project. Claremont, CA: Pomona College.

Hardin, B.D., G.P. Bond, M.R. Sitkov, F.D. Andrew, R.P. Beliles and R.W. Niemeier. 1981. Testing of selected workplace chemicals for teratogenic potential. Scand. J. Work Environ. Health. 7:66–75.

Harsch, D.E. and R.A. Rasmussen. 1977. Identification of methyl bromide in urban air. Anal. Let. 10:1041–1147.

Hawley, G.G. 1981. The Condensed Chemical Dictionary, 10th. New York, NY: Van Nostrand Reinhold Co. p. 670.

Hine, C.H. 1969. Methyl bromide poisoning. J. Occup. Med. 11(1):1–10. Data base abstr. from Health Aspects Pest. Abstr. Bull. 70:00079.

Honma, T., A. Sudo, M. Miyagawa, M. Sato and H. Hasegawa. 1982. Significant changes in monoamines in rat brain induced by exposure to methyl bromide. Neurobehav. Toxicol. Teratol. 4(5):521–524.

IARC. 1987. International Agency for Research on Cancer. IARC monographs on the evaluation of the carcinogenic risk of chemicals to humans. Supplement 7. Lyon, France: IARC, WHO. 7:245–246.

Irish, D.D., E.M. Adams, H.C. Spencer and V.K. Rose. 1940. The response attending exposure of laboratory animals to vapors of methyl bromide. J. Ind. Hyg. Toxicol. 22:218–230.

Ishitsu, S., H. Momotani, M. Sato, E. Nakayama and M. Minami. 1975. Methyl bromide intoxication. In: Proc. 18th Annual Meeting Japan Society of Industrial Medicine. pp. 62–63. Abstract.

Kool, H.J., C.F. van Kreijl and B.C.J. Zoeteman. 1982. Toxicology assessment of organic compounds in drinking water. Crit. Rev. Environ. Control. 12:307–347.

Kornbrust, D.J. and J.S. Bus. 1983. The role of glutathione and cytochrome P-450 in the metabolism of methyl chloride. Toxicol. Appl. Pharmacol. 67:246–256.

Kramers, P.G.N., C.E. Voogd, A.G.A.C. Knaap and C.A. van der Heijden. 1985. Mutagenicity of methyl bromide in a series of short-term tests. Mutat. Res. 155(1–2):41–47.

Longley, E.O. and A.T. Jones. 1965. Methyl bromide poisoning in man. Ind. Med. Surg. 34:499–502.

Longley, E.O. and G.R. Simpson. 1970. Acute methyl bromide poisoning. Proc. 4th Int. Congr. Rural Med. Data base abstr. from Health Aspects Pest. Abstr. Bull. 72:0507.

Lovelock, J.E. 1975. Natural halocarbons in the air and in the sea. Nature. 256:193–194.

Mabey, W. and T. Mill. 1978. Critical review of hydrolysis or organic compounds in water under environmental conditions. J. Phys. Chem. Ref. Data. 7:383–415.

Mabey, W.R., J.H. Smith, R.T. Podoll et al. 1981. Aquatic fate process data for organic priority pollutants. EPA 440/4-81-014. Washington, D.C.: Office of Water Regulations and Standards, U.S. EPA. (Cited in U.S. EPA, 1986b.)

McGregor, D.B. 1981. Tier II mutagenic screening of 13 NIOSH priority compounds: individual compound report: methyl bromide. NTIS PB83-130211. 190 pp. (Cited in U.S. EPA, 1986b.)

Medinsky, M.A., J.A. Bond, J.S. Dutcher and L.S. Birnbaum. 1984. Disposition of [^{14}C]-methyl bromide in Fischer 344 rats after oral or intraperitoneal administration. Toxicol. 32:187–196.

Medinsky, M.A., J.S. Dutcher, J.A. Bond et al. 1985. Uptake and excretion of [^{14}C]-methyl bromide as influenced by exposure concentration. Toxicol. Appl. Pharmacol. 78:215–225.

Meister, R., ed. 1988. Farm Chemicals Handbook. Willoughby, OH: Meister Publishing Company.

Merck Index. 1983. The Encyclopedia of Chemicals and Drugs, 9th ed. Rahway, NJ: Merck Co., Inc.

Miller, D.P. and H.W. Haggard. 1943. Intracellular penetration of bromide as a feature in the toxicity of alkyl bromides. Jr. Ind. Hyg. Toxicol. 25:423–433.

Moriya, M., T. Ohta, K. Watanabe, T. Miyazawa, K. Kato and Y. Shirasu. 1983. Further mutagenicity studies on pesticides in bacterial reversion assay systems. Mutat. Res. 116(3–4):185–216.

NTP. 1984. National Toxicology Program. No evidence of methyl bromide carcinogenicity found by NTP Panel. Pest. Tox. Chem. News. 13:9–10. November 28.

NTP. 1987. National Toxicology Program. Research and testing program. Management Status. Research Triangle Park, NC: NTP. October 13.

Perry, R.H. and C.H. Chilton. 1973. Chemical Engineers Handbook, 5th ed. New York, NY: McGraw Hill Book Co.

Ruth, J.H. 1986. Odor thresholds and irritation levels of several chemical substances: a review. Am. Ind. Hyg. Assoc. J. 47:A142-A151.

Shackelford, W.M. and L.H. Keith. 1976. Frequency of organic compounds identified in drinking water. EPA 600/4-76-062. Athens, GA: U.S. EPA.

Sikov, M.R., W.C. Cannon, D.B. Carr, R.A. Miller, L.F. Montgomery and D.W. Phelps. 1980. Teratologic assessment of butylene oxide, styrene oxide and methyl bromide. NTIS PB81-168510. 87 p.

Simmon, V.F. and R.G. Tardiff. 1978. The mutagenic activity of halogenated compounds found in chlorinated drinking water. In: Water Chlorination: Environmental Impact and Health Effects, Vol. 2. pp. 417-431.

Simmon, V.F., K. Kauhanen and R.G. Tardiff. 1977. Mutagenic activity of chemicals identified in drinking water. 2nd Int. Conf. Environmental Mutagens, Edinburgh, Scotland. July.

Singh, H.B., L.J. Salas and L.A. Cavanagh. 1977. Distribution, sources and sinks of atmospheric halogenated compounds. J. Air Pollut. Control Assoc. 27:332-336.

Singh, H.F., L.J. Salas and R.E. Stiles. 1983. Methyl halides in and over the eastern Pacific (40°N-32°S). J. Geophys. Res. 88:3684-3690.

Sittig, M. 1985. Handbook of Toxic and Hazardous Chemicals and Carcinogens, 2nd ed. Park Ridge, NJ: Noyes Publications.

Staples, C.A., A.F. Werner and T.J. Hoogheem. 1985. Assessment of priority pollutant concentrations in the United States using STORET Data Base. Environ. Toxicol. Chem. 4:131-142.

Stenger, V.A. 1978. Bromine compounds. In: Grayson, M. and D. Eckroth, eds. Kirk-Othmer Encyclopedia of Chemical Technology, Vol. 4, 3rd ed., New York, NY: John Wiley and Sons, Inc. pp. 251-252.

Torkelson, T.R., H.R. Hoyle and V.K. Rowe. 1966. Toxicological hazards and properties of commonly used space, structural and certain other fumigants. Pest. Control. 34(7):13-18, 40-42.

U.S. EPA. 1975. Preliminary assessment of suspected carcinogens in drinking water. Interim Report to Congress. Washington, DC: U.S. EPA, June 1975.

U.S. EPA. 1982. Tolerances and exemptions from tolerances for pesticide chemicals in or on raw agricultural commodities; inorganic bromides resulting from soil treatment with methyl bromide. Fed. Reg. 47:31550-31551.

U.S. EPA. 1983. Tolerances and exemptions from tolerances for pesticide chemicals in or on raw agricultural commodities; inorganic bromides resulting from fumigation with methyl bromide. Fed. Reg. 48:20052.

U.S. EPA. 1984. Memorandum from S. Schatzow, Director Office of Pesticide Programs to D.R. Clay, Director Office of Toxic Substances, FYI-OTS-1184–0327. Supplement, Sequence D. Washington, DC: Office of Pesticide Programs. November 9.

U.S. EPA. 1985a. Method 502.1 — Volatile halogenated organic compounds in water by purge and trap gas chromatography, Cincinnati, OH: Environmental Monitoring and Support Laboratory, June 1985 (Revised November 1985).

U.S. EPA. 1985b. Method 524.1 — Volatile organic compounds in water by purge and trap gas chromatography/mass spectrometry. Cincinnati, OH: Environmental Monitoring and Support Laboratory, June (Revised November 1985).

U.S. EPA. 1986a. Guidelines for carcinogen risk assessment. *Fed. Reg.* 51(185):33992–34003.

U.S. EPA. 1986b. Health and environmental effects profile for methyl bromide. Prepared by the Office of Health and Environmental Assessment, Environmental Criteria and Assessment Office, Cincinnati, OH for the Office of Solid Waste and Emergency Response, Washington, DC.

U.S. EPA. 1988. Intra-agency memorandum from Laurence D. Chitlik, Senior Science Advisor to Walter C. Francis, Registration Division. Methyl bromide rebuttal to letter of 10/19/87 from methyl bromide industry panel regarding their inhalation two generation reproduction study CASWEL #555, Project # 8–0927B. Office of Pesticide Programs, Washington, D.C., August 5.

Ushio, K. and R. Osozuka. 1977. A case of severe intoxication due to methyl bromide. Sangyo Igaku. 19:355–356. Abstract.

Verschueren, K. 1983. Handbook of Environmental Data on Organic Chemicals, 2nd ed. New York, NY: Van Nostrand Reinhold Co. pp. 835–836.

Voogd, C.E., A.G.A.C. Knaap, C.A. van der Heijden and P.G.N. Kramers. 1982. Genotoxicity of methylbromide in short-term assay systems. Mutat. Res. 97:233. Abstract.

Wong, O., W. Brocker, H.V. Davis and G.S. Nagle. 1984. Mortality of workers potentially exposed to organic and inorganic brominated chemicals. DBCP, TRIS, PBB and DDT. Br. J. Ind. Med. 4115–24. (Cited in U.S. EPA, 1986b.)

Worthing, C.R. and S.B. Walker, ed. 1983. The Pesticide Manual, 7th ed. Croydon, England: The British Crop Protection Council p. 372.

Zatuchni, J. and K. Hong. 1981. Methyl bromide poisoning seen initially as psychosis. Arch. Neurol. 38:529–530.

Dichlorodifluoromethane

I. General Information and Properties

A. CAS No. 75–71–8

B. Structural Formula

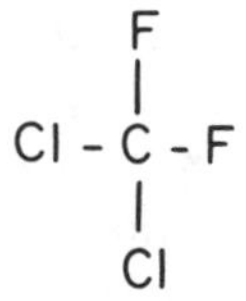

Dichlorodifluoromethane

C. Synonyms

- F-12; fluorocarbon 12; Freon-12.

D. Uses

- Applications of dichlorodifluoromethane (F-12) include use primarily as a refrigerant (60–80%) and blowing agent (15–25%), with minor use as a food freezant (3%) (WHO, 1986). Miscellaneous uses (1–10%) are as a leak detection agent, for chilling of cocktail glasses and as a low temperature solvent. Aerosol propellant use accounted for approximately 60% of production in 1972 but is now minor due to government restrictions (WHO, 1986).

E. Properties (Verschueren, 1983; WHO, 1986)

Chemical Formula	CCl_2F_2
Molecular Weight	120.92
Physical State (at 25°C)	Colorless gas
Boiling Point	–29.8°C
Melting Point	–111°C, –158°C
Vapor Pressure (20°C)	4,250 mm Hg
Specific Gravity (20°C)	1.329

Water Solubility (25°C)   280 mg/L
Log Octanol/Water
  Partition Coefficient   2.16
Taste Threshold   —
Odor Threshold   —
Conversion Factor   1 ppm = 4.95 mg/m^3
                             1 mg/m^3 = 0.202 ppm

F. Occurrence

- Singh et al. (1979) measured F-12 in the Pacific Ocean at surface concentrations of 0.13 ± 0.006 ng/L. Average concentrations at a 300 m depth were 0.06 ng/L.

- Singh et al. (1979) and Tyson et al. (1978) measured tropospheric levels of fluorocarbons in the northern and southern hemispheres. Mean atmospheric levels of 166–230 ppt (821–1,138 ng/m^3) for F-12 were found, with higher levels found in the northern hemisphere. F-12 has been monitored at the South Pole and in the U.S. Pacific Northwest (Rasmussen et al., 1981). In 1980, levels of these compounds were 284–322 ppt (1,405–1,594 ng/m^3). Over the period from 1976 to 1980, levels of F-12 increased at the rate of 9% per year at the South Pole.

- F-12 levels in Greenland Sea water (Bullister and Weiss, 1983) and in coastal waters (Tomita et al., 1983) in 1982 were 0.19–0.33 ng/L.

- F-12 has been detected in surface snow and rainwater in Alaska (Su and Goldberg, 1976).

G. Environmental Fate

- The dominant removal process of F-12 from water is most likely volatilization (WHO, 1986).

- The hydrolysis rate constant for F-12 at 30°C was too slow to be measured by an unspecified/analytical method as reported (DuPont de Nemours and Co., 1980).

- Su and Goldberg (1976) reported that F-12 is persistent in water, but specific information regarding biodegradation is not available. The volatility of F-12 might be expected to limit, if not preclude, biodegradation (WHO, 1986).

- Relatively little information is available pertaining specifically to the bioaccumulation of F-12. However, Neely et al. (1974) theorized that bioaccumulation is directly related to the log octanol/water partition coefficient of the compound. The experimentally determined log octanol/water partition coefficient for F-12 is reported to be 2.16 (Hansch and Leo, 1979), indicating a possibility for bioaccumulation under conditions of constant exposure. The high volatility of F-12 may be a more important factor in the environmental fate of this compound than partitioning.

II. Pharmacokinetics

A. Absorption

- F-12 would be expected to be absorbed from the gastrointestinal tract on the basis of its physical properties (low molecular weight and relatively high lipid solubility) (WHO, 1986). Quantitative data on the absorption of orally administered F-12 were not found in the available literature, but absorption following oral administration can be inferred from tissue F-12 levels and evidence of systemic toxicity.

B. Distribution

- Sherman (1974) studied tissue distribution of F-12 in male and female rats (Charles River — CD strain) and dogs (beagles) after 1 and 2 years of oral administration. There was some indication of storage in the adrenals, bone marrow and fat in both species, but the quantitative significance of these findings is difficult to assess (Sherman, 1974).

- Following 5-minute inhalation exposure of rats to 0.18–0.70% (8,900–35,000 mg/m³) F-12 in air, the highest tissue levels of F-12 were found in the adrenals, followed by the fat, and then the heart (Allen and Hansburys, Ltd., 1971). F-12 could be detected in the cerebrospinal fluid of dogs within the first 2.5 minutes of inhalation exposure to 20% (990 g/m³) F-12 (Paulet et al., 1975).

C. Metabolism

- Rats given an unspecified oral dose of ^{14}C-labeled F-12 metabolized approximately 2% of the dose to CO_2 (excreted in expired air) and excreted approximately 0.5% of the dose in the urine, presumably as unspecified metabolites (Eddy and Griffith, 1971).

- Blake and Mergner (1974) administered 8,200–11,800 ppm (40,000-58,400 mg/m³) of ^{14}C-labeled F-12 to anesthetized beagles for 12–20 minutes using an inhalation bag (total inhaled dose = 3,230-4,460 mg). Only approximately 1% of the administered radioactivity was recovered as ^{14}C-CO_2 and as nonvolatile urinary and tissue radioactivity, which represent metabolites. The impurities ($^{14}CF_3Cl$ and $^{14}CF_3$) in the ^{14}C-labeled F-12 (96% pure) were thought unlikely to account for the observed metabolites.

D. Excretion

- Administration of approximately 15 or 150 mg/kg bw/day of F-12 by gavage for 2 years to rats and 8 or 80 mg/kg bw/day of F-12 in food for 2 years to dogs did not result in significantly higher levels of fluoride in the urine by comparison with controls (Sherman, 1974). Considerable variation in fluoride excretion, however, was reported both within and among all groups throughout the test period. Other routes of excretion, such as expired air, were not monitored.

- Eddy and Griffith (1971) obtained results on the metabolism of ^{14}C-labeled F-12 in rats from oral administration showing a somewhat greater degree of metabolism. Approximately 2% of the total dose was exhaled as CO_2 and approximately 0.5% was excreted in the urine. By 30 hours after administration, neither F-12 nor metabolites were detected in body tissue as ^{14}C.

III. Health Effects

A. Humans

1. Short-term Exposure

- Data regarding the oral toxicity of F-12 to humans could not be located in the available literature.

- Stewart et al. (1978) monitored single exposures to F-12 by inhalation at concentrations of 0.025, 0.05 and 0.1% for 1 minute to 8 hours in a controlled-environment chamber, which induced no observable effects in groups of approximately 10 men and 10 women. Premature ventricular contractions were noted in one subject prior to a 1-hour exposure to F-12. However, the rate of the premature ventricular contraction was not affected by exposure to F-12 and these contractions were not observed in the subject 1 week later.

- Two male volunteers exposed by inhalation to 1,000 and 10,000 ppm (4,950 and 49,500 mg/m³) of F-12 for 2.5 hours had no adverse effects as monitored by subjective and clinical observations, laboratory tests and EKGs (Azar et al., 1972). During exposure to 10,000 ppm, however, their psychomotor test scores were 7% below their scores recorded during a nonexposed control period.

- Exposure of two subjects to 4% (200 g/m³) F-12 for 14–80 minutes was associated with a tingling sensation, humming in the ears, EEG changes, a feeling of apprehension, decreased psychological test scores and slurred speech. Exposure of one subject to 11% (540 g/m³) F-12 for 10 minutes produced a partial loss of consciousness with amnesia and a significant degree of cardiac arrythmia (Kehoe, 1943).

- A study by Marier et al. (1973) involved a group of 20 housewives who were given 13 household products containing fluorocarbon propellant (including F-12). The study was divided into three periods: (1) no use or light use of aerosol products for 2 weeks (pre-exposure period); (2) 4 weeks of heavy use defined as using 21.6 g of propellant/day/subject, which was estimated to be 2–25 times higher than normal use (exposure period); and (3) no use or light use of aerosol products for 2 weeks (post-exposure period). All women received physical examinations at the end of the first 2 weeks (pre-exposure), 6 weeks (exposure) and 8 weeks (post-exposure). The only effect noted was an increase in lactic dehydrogenase (LDH) during the exposure period, which nonetheless remained within normal limits. Because there was no accompanying increase in serum glutamic oxalacetic transaminase (SGOT), Marier et al. (1973) did not consider the increase in LDH to be significant.

2. Long-term Exposure

- Data quantifying the effects of longer-term exposure of humans to F-12 could not be located in the available literature.

B. Animals

1. Short-term Exposure

- No deaths occurred in rats administered a single oral dose of 1,000 mg/ kg of F-12 in peanut oil (Clayton, 1967). This was the maximum feasible single oral dose. No additional details were reported.

- Male rats administered 430 mg/kg/day of F-12 orally for 10 days showed no signs of toxicity (Clayton, 1967). Organ weights and the histological appearance of major organs, including the liver, were unaffected by F-12. Additional details were not provided.

- Sayers et al. (1930) exposed dogs, monkeys and guinea pigs by inhalation to F-12 at 20% (990 g/m³) for 7–8 hours/day for periods of 35–52 days in most cases. During the first couple of weeks, dogs and to a lesser extent guinea pigs developed tremors and ataxia during exposure. The subsidence of these effects seemed to indicate a tolerance to F-12 exposure. Also, during the first 2 or 3 weeks, a slight to moderate weight loss was noted along with an increase in red blood cell count and hemoglobin. Differential leukocyte count showed a slight decrease in lymphocytes and an increase in polymorphonuclear leukocytes. No variation from controls in frequency of pregnancy and bearing healthy young was noted in exposed guinea pigs.

2. Dermal/Ocular Effects

- Local chilling or freezing of tissue is a potential problem from liquid F-12 exposure to the skin or eye (Waritz, 1971).

- No dermal effects were seen in rabbits sprayed on the shaved skin with 40% F-12 in sesame oil daily for 12 applications (Scholz, 1962).

- Spray application of F-12 to the skin, tongue, soft palate and auditory canal of rats 1-2 times/day, 5 days/week for 5-6 weeks produced slight irritation of the skin and no remarkable effects on the other areas (Quevauviller et al., 1964; Quevauviller, 1965). Since F-12 is highly volatile, it is unlikely that spray application resulted in retention in tissue.

3. Long-term Exposure

- Clayton (1967) described 90-day feeding studies of the toxicity of F-12 to rats and dogs. Male and female rats received 160–379 mg/kg/day of F-12. F-12 treatment did not affect growth, behavior, hematological or serum glutamic pyruvic transaminase (SGPT) values or gross and microscopic appearance of organs and tissues, including the liver. A slight elevation of plasma alkaline phosphatase activity was noted in the females. Male and female dogs were administered 84–95 mg/kg/day. F-12 treatment produced no effects as judged by clinical examination, blood and urine tests or histological examination.

- Male and female rats were administered F-12 in corn oil for 2 years by intragastric intubation at calculated average doses of 15 and 150 mg/kg/day. Treatment began when the animals were exposed to the compound *in utero* (see Reproductive Effects). Except for a slight decrease in the rate of body weight gain by animals that received the higher dose level, no clinical signs of toxicity were observed. Oral administration of the test compound did not affect mortality nor did it alter any clinical laboratory measurements (hematology, urine analysis, liver function test). After 16 months, there was no histological evidence of toxicity attributable to the administration of F-12 (Sherman, 1974). Because of the decrease in body weight gain in females administered the high dose, the low dose of 15 mg/kg/day is considered the NOAEL for this study.

- Groups of four each male and female beagle dogs were fed F-12 at measured doses of 0, 8 or 80 mg/kg bw/day for 2 years (Sherman, 1974). No signs of toxicity or deaths occurred. Treated and control groups had no significant differences in food consumption; body weight; hematological; clinical chemistry and urinalysis values; organ weights; or histopathological findings. Urinary 17-ketosteroid excretion, an indicator of adrenal function, was also unaffected. A dose of 80 mg/kg/day thus is considered the NOAEL.

4. Reproductive Effects

- In the multigeneration reproductive and chronic toxicity study conducted by Sherman (1974), groups of 11 male and 21 female Charles River rats (F_0 generation) were administered the vehicle (corn oil) or F-12 by intragastric intubation at dose levels of approximately 15 and 150 mg/kg/day for 3 months and then mated. The administration of F-12 was stopped after day 18 or 19 of gestation and resumed on day 5 of lactation. This procedure was followed through 3 generations. The F_{3A} litters did not receive F-12 after weaning and were killed 4 weeks after weaning for histopathological examination. There were no treatment-related changes in fertility, gestation, viability or lactation indices in any generation or in the body weight gain and histopathological findings in the F_{3A} rats. A reproductive NOAEL of 150 mg/kg/day is identified.

5. Developmental Effects

- In a teratogenicity study conducted by Culik and Sherman (1973), groups of approximately 25 pregnant Charles River CD rats were administered F-12 in corn oil by intragastric intubation at dose levels of 0, 16.6 or 170.9 mg/kg/day on days 6–15 of gestation. F-12 did not affect food intake or body weight gain of the dams, or the number of implantation sites, resorptions and live fetuses/ litter. Embryonal and fetal development, as measured by body weight, crown-rump length and examination for gross external, skeletal and soft tissue abnormalities, were unaffected by treatment. A developmental NOAEL of 170 mg/kg/day is identified.

- In a study by Paulet et al. (1974), groups of 20 pregnant Wistar rats and 10 pregnant rabbits were exposed to a 20% (200,000 ppm or approximately 1 kg/m³) concentration of a mixture of F-11 and F-12 in air for 2 hours/day on days 4–16 of gestation (rat) or days 5–20 of gestation (rabbit). The proportion of F-12:F-11 in the mixture was 9:1. Half of the animals in each group were sacrificed at day 20 (rats) or day 30 (rabbits) of gestation and the rest carried to term. There were no treatment-related adverse effects on maternal or fetal body weights, number of implantations, resorptions, fetuses, stillbirths, weight of pups at birth and number of pups surviving at 1 and 4 weeks. No anomalies were observed in treated litters. The authors did not specify how the fetuses were examined.

6. *Mutagenicity*

- F-12 was not mutagenic to *Salmonella typhimurium* TA100 or TA1535 when tested in the presence or absence of a metabolic activating system (Longstaff et al., 1984) or to cultured Chinese hamster ovary cells (Krahn et al., 1982). Negative mutagenicity results were also obtained for a *Trandescantia* hybrid (Van't Hof and Schairer, 1982) and in a modified dominant lethal assay conducted in rats by intragastric intubation at approximately 15 or 150 mg/kg/day (Sherman, 1974).

7. *Carcinogenicity*

- In the multigeneration reproductive and chronic toxicity study described (Sherman, 1974), groups of 50 male and 50 female Charles River rats of the F_{1A} generation were administered average oral doses of 0, approximately 15 or approximately 150 mg/kg/day of F-12 in corn oil for up to 2 years. This treatment produced no evidence of carcinogenicity.

IV. Quantification of Toxicological Effects

A. One-day Health Advisory

The available data are insufficient to serve as the basis for the 10-kg child One-day HA. A single oral administration of the maximum feasible dose of 1,000 mg/kg of F-12 to rats produced no deaths in a study reported by Clayton (1967). No other parameters of toxicity were mentioned and no other pertinent studies were identified.

It is therefore recommended that the 10-kg child Ten-day HA of 40 mg/L (calculated below) be used as a conservative estimate for a One-day HA.

B. Ten-day Health Advisory

The study by Clayton (1967) has been selected to serve as the basis for the 10-kg child Ten-day HA. In this study, oral administration of F-12 to male rats at 430 mg/kg/day for 10 days produced no signs of toxicity or effects on the weights or histological appearance of major organs, including the liver.

The 10-day HA for the 10-kg child is calculated as follows:

$$\text{Ten-day HA} = \frac{(430 \text{ mg/kg/day}) (10 \text{ kg})}{(100) (1 \text{ L/day})} = 43 \text{ mg/L (rounded to } 40{,}000 \text{ } \mu g/L)$$

C. Longer-term Health Advisory

The 90-day feeding study in dogs by Clayton (1967) has been selected to serve as the basis for the Longer-term HA. In this study, oral administration of 160–379 mg/kg/day of F-12 to male and female rats produced no effects on growth, behavior, hematological values, SGPT activity and the gross and microscopic appearance of tissues and organs, including the liver. A slight elevation in plasma alkaline phosphatase level was noted in the females. Oral administration of 84–95 mg/kg/day of F-12 to dogs for 90 days produced no clinical signs of toxicity, and no effects were detected during blood and urine tests or histopathological examinations. Because the data for dogs provide a more clearly defined NOAEL in terms of dose and a lack of adverse effects, these data are chosen as the basis for the HA.

The Longer-term HA for a 10-kg child is calculated as follows:

$$\text{Longer-term HA} = \frac{(90 \text{ mg/kg/day}) (10 \text{ kg})}{(100) (1 \text{ L/day})} = 9 \text{ mg/L } (9{,}000 \text{ } \mu g/L)$$

The Longer-term HA for a 70-kg adult is calculated as follows:

$$\text{Longer-term HA} = \frac{(90 \text{ mg/kg/day}) (70 \text{ kg})}{(100) (2 \text{ L/day})} = 31.5 \text{ mg/L (rounded to } 30{,}000 \text{ } \mu g/L)$$

D. Lifetime Health Advisory

The 2-year oral study in rats by Sherman (1974) is selected as the basis for the Lifetime HA. It is the only available lifetime oral study measuring an end point other than reproductive effects. The 2-year exposure in dogs represents a lesser percentage of the animal lifetime than does the rat study, in which gavage administration of dosage levels of 15 or 150 mg/kg/day to male and female rats produced no histopathological effects or changes in survival. Body weight gains were depressed in the high-dose female rats. There was no change in food consumption. The lower dose, 15 mg/kg/day, is chosen as the basis for the DWEL.

Step 1: Determination of the Reference Dose (RfD)

$$\text{RfD} = \frac{(15 \text{ mg/kg/day})}{(100)} = 0.15 \text{ mg/kg/day}$$

Step 2: Determination of the Drinking Water Equivalent Level (DWEL)

$$\text{DWEL} = \frac{(0.15 \text{ mg/kg/day}) (70 \text{ kg})}{(2 \text{ L/day})} = 5.25 \text{ mg/L (rounded to } 5{,}000 \text{ } \mu g/L)$$

Step 3: Determination of the Lifetime Health Advisory

Lifetime HA = (5.25) (0.2) = 1.05 mg/L (rounded to 1,000 µg/L)

E. Evaluation of Carcinogenic Potential

- The chronic gavage study by Sherman (1974) was negative for the carcinogenicity of F-12 to Charles River rats. No other carcinogenicity studies of F-12 were found in the available literature.

- IARC has not evaluated the carcinogenic potential of F-12. Applying the criteria described in the U.S. EPA's Guidelines for Carcinogen Risk Assessment (U.S. EPA, 1986c), F-12 may be classified in Group D: not classifiable. This category is for agents with inadequate animal evidence of carcinogenicity.

V. Other Criteria, Guidance and Standards

- U.S. EPA (1982) recommended an ambient water quality criterion of 28 mg/L based on an ADI of 56 mg/day for a 70-kg man (0.8 mg/kg/day). The ADI was calculated from a chronic (2-year) oral study for a dog NOAEL of 80 mg/kg/day (Sherman, 1974) and an uncertainty factor of 100 (U.S. EPA, 1982).

- The NAS (1980) Suggested-No-Adverse-Response Level (SNARL) for 1-day exposure of a 70-kg adult to F-12 is 350 mg/L based on an oral nonlethal dose for rats of 1,000 mg/day (Clayton, 1967) and an uncertainty factor of 100. The SNARL for 7-day exposure of a 70-kg adult is 150 mg/L based on a rat NOAEL of 430 mg/kg/day (Clayton, 1967) and an uncertainty factor of 100. The SNARL for chronic exposure is 5.6 mg/L based on a NOAEL of 80 mg/kg/day in dogs (Sherman, 1974) and an uncertainty factor of 100 (NAS, 1980).

- The ACGIH (1985) has adopted a Time-weighted Average Threshold Limit Value (TWA-TLV) of 1,000 ppm (approximately 4,950 mg/m^3) for occupational exposure to F-12. The current OSHA standard (PEL) for occupational exposure to F-12 is 1,000 ppm (Code of Federal Regulations, 1985).

- The New York State Department of Health (1989) adopted an MCL of 0.005 mg/L (5 µg/L) for dichlorodifluoromethane, effective January 9, 1989 (personal communication).

VI. Analytical Methods

- Analysis of F-12 is by a purge-and-trap gas chromatographic procedure used for the determination of volatile organohalides in drinking water (U.S. EPA, 1984). This method consists of bubbling an inert gas through the sample and trapping F-12 on an adsorbent material. The adsorbent material is then heated to release the F-12 onto a gas chromatographic column. This method is applicable to the measurement of F-12 over a concentration range of 0.03–1,500 μg/L. Confirmatory analysis is by mass spectrometry (U.S. EPA, 1985). The detection limit for confirmation by mass spectrometry is 0.33 μg/L.

VII. Treatment Technologies

- The available information suggests that F-12 is amenable to removal from water by air stripping.

- Methodology that consists of comparing the mass transfer rate constants of a number of compounds, including F-12, with one another and with that of oxygen under conditions of controlled energy input, was demonstrated by Roberts and Dandliker (1982). The rate of mass transfer of the compounds under investigation was shown to be proportional to that of oxygen, with a proportionality coefficient of approximately 0.6. In the turbulent region, the transfer rate constants of the chlorinated compounds as well as oxygen were proportional to the power input. No removal data are provided.

- Air stripping experiments were conducted in a column of 8.66 inch diameter packed with 1/2-inch ceramic Berl saddles to a height varying from 7.9–55 inches. The contaminants' influent concentrations were in the range of 50–200 μg/L. Three semi-empirical mass transfer models (Onda, Shulman and Sherwood-Holloway) were tested for their ability to predict gas stripping of trace organic contaminants from aqueous solutions. The Onda model provides the best overall fit (Roberts et al., 1985).

- U.S. EPA (1986a) estimated the feasibility of removing F-12 from water by air stripping employing the engineering design procedure and cost model presented at the 1983 National ASCE Conference on Environmental Engineering. Based on chemical and physical properties and assumed operating conditions, 99% removal efficiency of F-12 was reported by a column with a diameter of 5.5 feet and packed with 18.5 feet of 1-inch plastic saddles. The air-to-water ratio required to achieve this degree of removal effectiveness is 3:1.

- The amenability of F-12 to aeration has been clearly established. Actual system performance data, however, are necessary to determine the feasibility of using air stripping for the removal of F-12 from individual contaminated drinking water supplies.

VIII. References

ACGIH. 1985. American Conference of Governmental Industrial Hygienists. TLVs—Threshold limit values for chemical substances in the work environment, adopted by ACGIH with intended changes for 1985–1986. ACGIH, Cincinnati, OH.

Allen and Hansburys, Ltd. 1971. An investigation of possible cardiotoxic effects of the aerosol propellants, arctons 11 and 12, Vol. 1. Unpublished report, courtesy of D. Jack, Managing Director, Allen and Hansburys, Ltd. (Cited in WHO, 1986.)

Azar, A., C.F. Reinhardt, M.E. Maxfield, P.E. Smith, Jr. and L.S. Mullin. 1972. Experimental human exposures to fluorocarbon 12 (dichlorodifluoromethane). Am. Ind. Hyg. Assoc. J. 33(4):207–216.

Blake, D.A. and G.W. Mergner. 1974. Inhalation studies on the biotransformation and elimination of [^{14}C]trichlorofluoromethane and [^{14}C]dichlorodifluoromethane in beagles. Toxicol. Appl. Pharmacol. 30(3):396–407.

Bullister, J.L. and R.F. Weiss. 1983. Anthropogenic chlorofluoromethanes in the Greenland and Norwegian seas. Science. 221:265–268.

Clayton, W.J., Jr. 1967. Fluorocarbon toxicity and biological action. Fluorine Chem. Rev. 1:197–252.

Code of Federal Regulations. 1985. OSHA Occupational Standards Permissible Exposure Limits. 29 CFR 1910.1000.

Culik, R. and H. Sherman. 1973. Teratogenic study in rats with dichlorodifluoromethane (Freon 12). Medical Research Proj. No. 1388. Haskell Laboratory Report No. 206–73. Unpublished, Courtesy of DuPont de Nemours, Inc. 10 p.

DuPont de Nemours and Co. 1980. Freon Products Information Booklet B-2 A98825, 12/80. DuPont de Nemours and Co., Wilmington, DE. (Cited in WHO, 1986.)

Eddy, C.W. and F.D. Griffith. 1971. Metabolism of dichlorodifluoromethane C^{14} by rats. Presented in 1971 American Industrial Hygiene Assoc. Conf., Toronto, Canada, May. (Cited in Waritz, 1971.)

Hansch, C. and A.J. Leo. 1979. Substituent Constants for Correlation Analysis in Chemistry and Biology. New York, NY: John Wiley and Sons, Inc.

Kehoe, R.A. 1943. Report on human exposure to dichlorodifluoromethane in air. Unpublished report. Kettering Laboratory, University of Cincinnati, Cincinnati, OH. (Cited in Azar et al., 1972.)

Krahn, D.F., F.C. Barsky and K.T. McCooey. 1982. CHO/HGPRT mutation assay: evaluation of gases and volatile liquids. In: Genotoxic Effects of Airborne Agents. Environ. Sci. Res. 25:91-103.

Longstaff, E., M. Robinson, C. Bradbrook, J.S. Styles and I.F.H. Purchase. 1984. Genotoxicity and carcinogenicity of fluorocarbons: assessment by short-term *in vitro* tests and chronic exposure in rats. Toxicol. Appl. Pharmacol. 72:15-31.

Marier, G., G.H. McFarland and P. Dussault. 1973. A study of blood fluorocarbon levels following exposure to a variety of household aerosols. Household Pers. Prod. Ind. 10:68, 70, 92, 99. (Cited in WHO, 1986.)

NAS. 1980. National Academy of Sciences. Drinking Water and Health, Vol. 3. NAS, Washington, DC. pp. 101-104.

Neely, W.B., D.R. Branson and G.E. Blau. 1974. Partition coefficient to measure bioconcentration potential of organic chemicals in fish. Environ. Sci. Technol. 8:1113-1115. (Cited in WHO, 1986.)

New York State Department of Health. 1989. Personal communication to James Murphy, Office of Drinking Water, U.S. Environmental Protection Agency.

Paulet, G., S. Desbrousses and E. Vidal. 1974. *Absence d'effet teratogene des fluorocarbones chez le rat et le lapin.* [Absence of teratogenic effects of fluorocarbons in the rat and rabbit.] Arch. Mal. Prof. Med. Trav. Secur. Soc. 35:658-662. (Fre.)

Paulet, G., J. Lanoe, A. Thos, P. Toulouse and J. Dassonville. 1975. Fate of fluorocarbons in the dog and rabbit after inhalation. Toxicol. Appl. Pharmacol. 34(2):204-213.

Quevauviller, A. 1965. *Hygiene et securite des puseurs pour aerosols medicamenteux.* Prod. Probl. Pharm. 20(1):14-29. (Fre.)

Quevauviller, A., M. Schrenzel and H. Vu-Ngoc. 1964. *Tolerance locale (peau, muqueuses, plaies, brulures) chez l'animal, aux hydrocarbures chlorofluores* [Local tolerance in animals to chlorofluorinated hydrocarbons]. Therapie. 19:247-263. (Fre.)

Rasmussen, R.A. et al. 1981. Atmospheric trace gases in Antarctica. Science. 211:285-287. (Cited in WHO, 1986.)

Roberts, P.V. 1984. Comment on mass transfer of volatile organic contaminants from aqueous solution to the atmosphere during surface aeration. Correspondence. Environ. Sci. Technol. 18:890.

Roberts, P.V. and P. Dandliker. 1982. Mass transfer of organic contaminants during surface aeration. Am. Water Works Assoc. Ann. Conf. pp. 755-763.

Roberts, P.V. et al. 1984. Volatilization of organic pollutants in wastewater treatment: model studies. EPA 600/2-84-047.

Roberts, P.V., G.D. Hopkins, C. Munz and A.H. Riogas. 1985. Evaluating two-resistance models for air stripping of volatile organic contaminants in a countercurrent packed column. Environ. Sci. Technol. 19(23):164-173.

Sayers, R.R., W.P. Yant, J. Chornyak and H.W. Shoaf. 1930. Toxicity of dichlorodifluoromethane: a new refrigerant. U.S. Bureau of Mines Report, R.I. 3013. May. (Cited in WHO, 1986.)

Scholz, J. 1962. New toxicologic investigation of freons used as propellants for aerosols and sprays. Fortschr. Biol. Aerosol Forsch. 1957-1961. Ber. Aerosol-Kongr. 4:420-429. (Cited in Waritz, 1971.)

Sherman, H. 1974. Long-term feeding studies in rats and dogs with dichlorodifluoromethane (Freon 12 Food Freezant). Medical Research Proj. No. 1388. Haskell Laboratory Report No. 24-74. Unpublished. Courtesy of DuPont de Nemours and Co.

Singh, H.B., L.J. Salas, H. Shigeishi and E. Scribner. 1979. Atmospheric halocarbons, hydrocarbons, and sulfur hexafluoride: global distributions, sources, and sinks. Science. 203:899-903.

Stewart, R.D., P.E. Newton, E.D. Baretta, A.A. Herrmann, H.V. Forster and R.J. Soto. 1978. Physiological response to aerosol propellants. Environ. Health Perspect. 26:275-285.

Su, C.W. and E.D. Goldberg. 1976. Environmental concentrations and fluxes of some halocarbons. In: H.L. Windom and R.A. Due, ed. Marine Pollutant Transfer. Lexington, MA: Lexington Books; D.C. Heath and Company. pp. 353-374. (Cited in WHO, 1986.)

Tomita, I., S. Saitou, M. Meguro and H. Kanamori. 1983. Studies on the photolysis of trichlorofluoromethane and dichlorodifluoromethane in water. III. Determination of dichlorodifluoromethane in coastal water. Eisei Kagaku. 29(2):76-82. (Jap.) (CA 99(12):93392m)

Tyson, B.J. et al. 1978. Interhemispheric gradients of CF_2Cl_2, $CFCL_3$, CCl_4 and H_2O. Geophys. Res. Lett. 5:535-538. (Cited in WHO, 1986.)

U.S. EPA. 1982. U.S. Environmental Protection Agency. Errata: halomethanes. Ambient Water Quality Criterion for the Protection of Human Health. Prepared by the Office of Health and Environmental Assessment, Environmental Criteria and Assessment Office, Cincinnati, OH, for the Office of Water Regulations and Standards, Washington, DC.

U.S. EPA. 1984. U.S. Environmental Protection Agency. Method 601 - Purgeable halocarbons. *Fed. Reg.* 49(209):29-39.

U.S. EPA. 1985. U.S. Environmental Protection Agency. Method 524.1 - Volatile organic compounds in water by purge and trap gas chromatography/mass spectrometry. Environmental Monitoring and Support Laboratory, Cincinnati, OH. June (Revised November, 1985).

U.S. EPA. 1986a. U.S. Environmental Protection Agency. Economic evaluation of dichlorodifluoromethane removal from water by packed column air stripping. Prepared by the Office of Water for Health Advisory Treatment Summaries.

U.S. EPA. 1986b. U.S. Environmental Protection Agency. Integrated Risk Information System (IRIS). Reference dose (RfD) for oral exposure for dichlorodifluoromethane (verification date 07/22/85). Office of Health and Environmental Assessment, Environmental Criteria and Assessment Office, Cincinnati, OH.

U.S. EPA. 1986c. Guidelines for Carcinogen Risk Assessment. *Fed. Reg.* 51(185):33992–34003.

Van't Hof, J. and L.A. Schairer. 1982. *Trandescantia* assay system for gaseous mutagens. A report of the U.S. Environmental Protection Agency Gene-Tox Program. Mutat. Res. 99(3):303–315.

Verschueren, K. 1983. Handbook of Environmental Data on Organic Chemicals, 2nd ed. New York, NY: Van Nostrand Reinhold Co.

Waritz, R.S. 1971. Toxicology of some commercial fluorocarbons. Aerospace Med. Res. Lab., Wright-Patterson Air Force Base, Dayton, OH. NTIS AD751-429. (Microfiche)

WHO. 1986. World Health Organization. Environmental health criteria: chlorofluorocarbons. Developmental Draft Document. Prepared by the Office of Health and Environmental Assessment, Environmental Criteria and Assessment Office, U.S. EPA, Cincinnati, OH, for the World Health Organization.

Naphthalene

I. General Information and Properties

A. CAS No. 91–20–3

B. Structural Formula

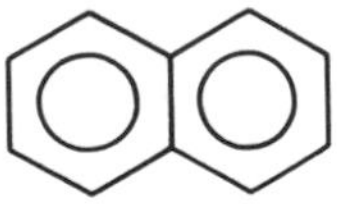

Naphthalene

C. Synonyms

- Moth balls, camphor tar, naphthene, albo carbon, naphthalin, moth flake, white tar (Chemline, 1988).

D. Uses

- Naphthalene is used in the manufacture of phthalic and anthranilic acids and other derivatives, and in making dyes; in the manufacture of resins, celluloid, lampblack and smokeless gunpowder; and as moth repellant, insecticide, anthelmintic, vermicide and intestinal antiseptic (U.S. EPA, 1980).

E. Properties (Hine and Mookerjee, 1975; Kenaga, 1980; Amoore and Hautala, 1983; Garten and Trabalka, 1983; Verschueren, 1983; Windholz, 1983; Hansch and Leo, 1985)

Chemical Formula	$C_{10}H_8$
Molecular Weight	128.16
Physical State (at 25°C)	White solid
Boiling Point	217.9°C (sublimes)
Melting Point	80.2°C
Density (20°C)	1.162 g/mL
Vapor Pressure (25°C)	0.082 mm Hg

Specific Gravity (20°C)   1.162
Water Solubility (25°C)   31.7 mg/L
Log Octanol/Water   3.01–4.70
 Partition Coefficient
Taste Threshold)   —
Odor Threshold (water)   0.021 mg/L (w/v)
Odor Threshold (air)   0.084 ppm (v/v)
Conversion Factor   1 mg/m^3 = 0.191 ppm
                                     1 ppm = 5.24 mg/m^3

F. Occurrence

- Naphthalene was not detected in drinking water (tap) from two midwestern U.S. cities (detection limit = 10 μg/L) (Callahan et al., 1979). Naphthalene, however, was detected at a concentration of 1 μg/L in the District of Columbia (Scheiman et al., 1974), 4.8 to 6.8 ng/L in Ottawa, Ontario, Canada (Benoit et al., 1979); 2.2 μg/L in Kitakyushu, Japan (Shinohara et al., 1981); and 1.2 to 8.8 ng/L in Norway (Kveseth et al., 1982).

- Staples et al. (1985) determined a median concentration of < 10 μg/L naphthalene in 1,206 industrial effluent samples recorded in the U.S. EPA's STORET data base during the years 1980 through 1983. Only 1.6% of the samples contained detectable levels (detection limit = 10 μg/L). U.S. EPA (1986a) summarized data regarding the concentration of naphthalene in ground water near waste sites and other likely contamination sites and reported a level of 110 μg/L naphthalene associated with a sanitary landfill and 1,800 μg/L in ground water near an underground coal gasification site in northeastern Wyoming.

- Reported concentrations of naphthalene in surface waters include 6 to 10 μg/L in a river (not identified) near a chemical production plant (Jungclaus et al., 1978) and 0.7 to 0.9 μg/L in the Delaware River near Philadelphia (Sheldon and Hites, 1978, 1979). Cole et al. (1984) detected naphthalene at average concentrations of 0.8 to 2.3 μg/L at a frequency of 11% in runoff water from several urban areas.

- U.S. EPA (1982) reported average values of 180 and 400 ppt (940 and 2,100 ng/m^3) from several U.S. urban/suburban areas and source dominated areas, respectively. Jarke et al. (1981) determined that the frequency of detecting naphthalene in the air within Chicago area houses (43%) was approximately twice as high as the frequency in outdoor air (21%).

- According to Galloway et al. (1983), naphthalene was detected in food, notably marine mussels from the East, West and Gulf Coasts at a range of 2.8 µg/kg dry weight (±0.8 s.d.) to 96 µg/kg dry weight (±118 s.d.). Pellizzari et al. (1982) detected naphthalene in 6/8 samples (concentration not reported) of human milk collected in urban areas of New Jersey, Pennsylvania and Louisiana.

G. Environmental Fate

- The half-life of naphthalene at a concentration of 0.03 µg/L in the Rhine River was experimentally determined to be 2.3 days (Zoeteman et al., 1980).

- Mackay and Leinonen (1975) estimated a half-life of 7.2 hours for the volatilization of naphthalene (quantity not stated) from an aqueous solution 1 meter deep, based on the liquid-phase mass transfer coefficient. Southworth (1979) estimated that a 10-fold increase in current velocity would accelerate volatilization two to three times. Based on the Exposure Analysis Modeling System (EXAMS) technique, Rodgers et al. (1983) estimated a first-order volatilization rate constant of 0.16 hour^{-1}, which resulted in a half-life of 4.3 hours (U.S. EPA, 1986a).

- Lee et al. (1978) reported that sedimentation removed 11% of the total naphthalene present in crude oil in a controlled aquatic ecosystem enclosure. Up to 5% of the naphthalene was subject to microbial degradation each day.

- Biodegradation and volatilization from soil surfaces are the two important processes for the removal of naphthalene from soil (U.S. EPA, 1986a). The estimated volatilization half-lives for naphthalene from a soil containing 1.25% organic carbon were 1.1 day from soil 1 cm deep and 14 days from soil 10 cm deep (Jurg et al., 1984). The overall half-life of naphthalene in soil from a waste disposal site was estimated to be 3.6 months (Zoeteman et al., 1981).

- Biodegradation is enhanced by microbial acclimation and by the presence of nutrients in water and is temperature- and concentration-dependent (U.S. EPA, 1986a). Maximum biodegradation occurs at pH 8 and in the presence of a positive redox potential (U.S. EPA, 1986a).

- Data on the photolysis of naphthalene and its reaction rate with singlet oxygen ($^1O_2$) and peroxy radical ($RO_2 \cdot$) suggest that these processes are not likely to have a significant impact on the fate of naphthalene in water (U.S. EPA, 1986a).

II. Pharmacokinetics

A. Absorption

- Quantitative data regarding the gastrointestinal, pulmonary or dermal absorption of naphthalene could not be located in the available literature. Reports from the literature indicate qualitatively that pulmonary and gastrointestinal absorption occurs. Naphthalene can be sufficiently absorbed when ingested as a solid to result in toxicity in humans (U.S. EPA, 1986a).

- Humans most absorb naphthalene by the inhalation route (Sandmeyer, 1981). Valaes et al. (1963) reported toxicity and death in newborn infants exposed to naphthalene vapors from clothes or blankets that had been stored in or near the infant's room.

- Cutaneous and gastrointestinal absorption are facilitated when naphthalene is administered with oil or fat, respectively (Sandmeyer, 1981).

B. Distribution

- Eisele (1985) administered ^{14}C-naphthalene orally to animals raised for food (laying hens, swine and dairy cattle). This was either as a single exposure with sacrifice 1 or 3 days later or on a daily basis for 31 days with sacrifice on day 32. In these species, the highest levels were mostly found in liver, lung, kidney and fat. Hens treated by both regimens were found to have naphthalene distributed to the following tissues in highest to lowest concentrations: kidney > lungs > liver and fat. Distribution to fat was lower after chronic exposure. In swine, levels were highest in the fat and kidneys following a single dose; however, after 31 days of treatment, highest levels were in lung, liver and heart, with only low levels in the fat and kidney. The liver and muscle were major deposition sites in dairy cattle. Radioactivity was found in eggs from treated hens (primarily in the yolk) and was also found in milk from treated cows.

- Naphthalene and its metabolites have been reported to cross the human placenta in amounts sufficient to cause fetal toxicity (Zinkham and Childs, 1958; Anziulewicz et al., 1959).

C. Metabolism

- *In vitro* studies of naphthalene indicate that oxidation to the epoxide, naphthalene 1,2-oxide, is the initial biotransformation reaction in rats (Jerina et al., 1970). This intermediate may then be converted to a number of other oxidation products (e.g., phenols, dihydrodiols) or be conjugated with glutathione (Boyd et al., 1972; Chen and Dorough, 1979).

- Chen and Dorough (1979) administered a single 100 mg/kg intraperitoneal (i.p.) dose of ^{14}C-naphthalene in a Tween 80 and ethanol suspension to adult female Sprague-Dawley rats (200 g) and quantified the urinary metabolites. Urinary radioactivity (collected for 72 hours) accounted for 60% of the administered dose. The ether-extractable portion of the urine accounted for 6% of the administered dose and consisted primarily of 1-naphthol and 1,2-dihydro-1,2-dihydroxynaphthalene at 60 and 28%, respectively, of the ether-extractable radioactivity. Water soluble metabolites included 1-naphthol; 1,2-dihydro-1,2-dihydroxy-1-naphthyl sulfate; 1,2-dihydro-2-hydroxy-1-naphthyl glucuronide and N-acetyl-S-(1,2-dihydro-2-hydroxy-1-naphthyl) cysteine at 5.0, 8.0, 16.8 and 65.0% of the nonether-extractable urinary radioactivity, respectively. These authors concluded that glutathione and mercapturic acid conjugation are major detoxification pathways in rats.

- Summer et al. (1979) administered single gavage doses of 0, 30, 75 or 200 mg/kg naphthalene in sesame or corn oil to adult male SPF Wistar rats (200 to 250 g) and to male and female chimpanzees (12–17 kg). A dose-related increase in the 24-hour urinary excretion of thioethers was observed in the rats. Doses ≤ 200 mg/kg did not result in increased urinary excretion of thioethers in chimpanzees. The authors reported urinary thioether excretion rates in untreated rats, chimpanzees and humans of 94.4, 18.0 and 22.8 μmol/24 hours/kg, respectively. Noting the similarities with regard to urinary excretion of thioethers among chimpanzees and humans and the differences among primates and rats, these authors suggested that species differences in the activity of glutathione-S-transferase result in substantial glutathione conjugation in the rat and a negligible amount in primates. Biliary excretion of glucuronide conjugate products was not investigated in this experiment.

- Several investigators reported that reactive metabolites bind irreversibly with eye lens proteins and are associated with cataract formation. They also bind covalently with macromolecules in the lungs and may be associated with damage to the bronchiolar epithelium (U.S. EPA, 1986a).

D. Excretion

- Chen and Dorough (1979) administered a single 100 mg/kg i.p. dose of ^{14}C-naphthalene to adult female Sprague-Dawley rats and collected urine and feces for 72 hours. By 24 hours, urinary radioactivity accounted for 24% of the administered dose. Total recovery of radioactivity at the end of the 72-hour collection period was 74% of the administered dose, 60% in the urine and 14% in the feces.

- Summer et al. (1979) administered single gavage doses of naphthalene of 0, 30, 75 or 200 mg/kg to two male and two female yearling chimpanzees and five adult male SPF Wistar rats, and determined urinary excretion of thioether. At 0, 30, 75 and 200 mg/kg, thioether excretion in rats was 94.4, 185.6, 279.6 and 502.0 μmol/24 hours/kg, respectively. Thioether excretion by chimpanzees (measured at 0 and 200 mg/kg) did not increase as a result of exposure to naphthalene.

III. Health Effects

A. Humans

1. Short-term Exposure

- Bilateral cataracts and optic atrophy were diagnosed in a 36-year-old man, 1 year after he ingested 5 g of naphthalene in oil (Lezenius, 1902). He was nearly blind within 9 hours of ingesting the naphthalene.

- There are numerous case reports of hemolytic anemia, accompanied by jaundice and occasionally by renal disease, occurring in infants, children and adults exposed to unknown levels of naphthalene by ingestion, inhalation or dermal contact. No exposure levels were reported (U.S. EPA, 1980).

- Upon direct skin contact, naphthalene is a primary irritant (Sandmeyer, 1981). Diapers or clothes stored with mothballs and used directly on infants have caused skin rashes and systemic poisoning (Sandmeyer, 1981).

- Dreisbach and Robertson (1987) reported a fatal dose from oral exposure to be approximately 2 g. No details were provided.

2. Long-term Exposure

- Van der Hoeve (1906) reported the development of cataracts and retinal hemorrhage in a 44-year-old man occupationally exposed to powdered naphthalene. A coworker developed unilateral chorioretinitis. Exposure levels were not reported.

- Ghetti and Mariana (1956) reported that 8/21 workers exposed to naphthalene fumes or dust for ≤ 5 years in a manufacturing setting developed cataracts. This finding was attributed to naphthalene exposure because the workers were ≤ 50 years old and not likely to develop spontaneous cataracts. Levels of exposure were not known.

- Toxic effects in infants have been associated with naphthalene exposure (level not reported) of the mother during gestation (Zinkham and Childs, 1958; Anziulewicz et al., 1959). Infants and adults with congenital erythrocyte glucose 6-phosphate dehydrogenase deficiency are more susceptible to naphthalene-induced hemolytic anemia (U.S. EPA, 1986a).

B. Animals

1. Short-term Exposure

- Shopp et al. (1984) administered single gavage doses of 0, 200, 400, 600, 800 or 1,000 mg/kg naphthalene in corn oil to groups of eight male and eight female CD-1 mice. LD_{50} values were 533 mg/kg for males and 710 mg/kg for females. Decedents and survivors of a 14-day observation period were necropsied. All mice, except males exposed to 200 mg/kg and females exposed to ≤ 400 mg/kg, developed ocular ptosis and red discharge within 1 hour of treatment.

- Gaines (1969) reported LD_{50} values of 2,200 mg/kg for male Sherman rats and 2,400 mg/kg for female Sherman rats given oral doses of naphthalene. Other oral LD_{50} values for rats (strains not specified) exposed to naphthalene include 1,780 mg/kg (NIOSH, 1977) and 9,430 mg/kg (Union Carbide Corp., 1968). No explanation has been offered for the wide range in values.

- Zuelzer and Apt (1949) administered 420 and 1,530 mg/kg naphthalene (in a solid form) in a single oral dose to dogs and observed decreases of 29 and 33%, respectively, in blood hemoglobin concentrations.

- Shopp et al. (1984) intubated groups of 40 to 76 female and 76 to 112 male CD-1 mice with concentrations of 0, 27, 53 or 267 mg/kg/day naphthalene in corn oil for 14 consecutive days. Adverse effects, associated only with the 267 mg/kg/day concentration, included increased mortality and decreased terminal body weights in male and female mice, decreased absolute thymus weight (30%) in males, and increased bilirubin and decreased absolute and relative spleen and lung weights in females. There were no effects on hexobarbital sleeping time or on various immunological screening tests, with the exception that high-dose females had decreased response to concanavalin A in lymphocytes. A NOAEL of 53 mg/kg was identified. The authors suggest that the lack of naphthalene-induced hemolytic anemia or cataract formation may have been due to characteristics of the CD-1 strain.

2. Dermal/Ocular Effects

- Van Heyningen and Pirie (1976) administered 1,000 mg/kg/day naphthalene by gavage to rabbits and observed cataract formation and retinal changes as early as 1 and 2 days.

- Shopp et al. (1984) administered single gavage doses of 0, 200, 400, 600, 800 or 1,000 mg/kg naphthalene in corn oil to groups of eight male and eight female CD-1 mice. Ocular ptosis and red discharge were observed within 1 hour of treatment in all mice, except males exposed to 200 mg/kg and females exposed to ≤400 mg/kg.

3. Long-term Exposure

- NTP (1980a) administered gavage doses of 0 (corn oil vehicle), 25, 50, 100, 200 or 400 mg/kg/day naphthalene, 5 days/week for 13 weeks to groups of 10 male and 10 female F344 rats. Frank signs of toxicity and the death of two males at the 400 mg/kg dose were noted in the first week. Decreases in body weight gain (> 10%) occurred in male rats exposed to ≥200 mg/kg and in female rats exposed to ≥100 mg/kg. Hematological effects, observed at 400 mg/kg but not at lower doses, included a marginal decrease in hemoglobin and hematocrit in both sexes and a slightly altered differential white blood cell count in males. Comprehensive histopathological examinations were performed on control and high-dose rats. Lymphoid depletion of the thymus occurred in females exposed to 400 mg/kg, and renal lesions occurred in males exposed to ≥200 mg/kg. Diffuse tubule degeneration was observed in one male exposed to 400 mg/kg. Mild lymphocytic infiltration and tubule regeneration were observed in two males exposed to 200 mg/kg. A NOAEL was observed to be 50 mg/kg (5 days/week).

- In a similar study, NTP (1980b) administered 0 (corn oil vehicle), 12.5, 25, 50, 100 or 200 mg/kg/day naphthalene by gavage to groups of 10 male and 10 female B6C3F$_1$ mice 5 days/week for 13 weeks. No treatment-related deaths occurred. Transient signs of toxicity (lethargy, roughened haircoats and decreased food consumption) which occurred in both males and females exposed to 200 mg/kg were not attributed to treatment. The rate of body weight gain in all groups of treated males was slightly greater than that of controls but was not dose related. Hematological evaluations, performed on all groups, and comprehensive histopathological examinations, performed on control and high-dose groups, revealed no compound-related effects. A NOAEL of 200 mg/kg/day (5 days/week) was thus identified in this study.

- Shopp et al. (1984), administered 0, 5.3, 53 or 133 mg/kg/day naphthalene by gavage to 6-week-old female (40 to 76) and male (76 to 112) CD-1 mice for 90 consecutive days. Untreated and vehicle (corn oil) treated controls were maintained. There were no compound-related effects on mortality or body weights at 90 days. Absolute and relative spleen weights and absolute brain and liver weights were significantly reduced in high-dose females. There were no adverse observations in gross pathology or several immunological screening tests. Small changes in several serum chemistry parameters (i.e., blood protein, glucose, creatine, blood urea nitrogen, calcium ion) were observed, mostly at the middle and high doses. These changes were considered to possibly be treatment related adverse effects. A significant dose-related inhibition of aryl hydrocarbon hydroxylase activity was observed in both sexes in all treated groups, but the significance of this finding to health is not clear. Due to the uncertainty regarding the serum chemistry changes, a NOAEL could not be identified from this study.

- Schmahl (1955) administered 10 to 20 mg/day naphthalene in the diet to a total of 28 BDI and BDIII rats, 6 days/week for 700 days. The rats were about 100 days old at the start of the experiment. Treatment was ended after 700 days, at which time a total of 10 g naphthalene/rat had been administered. The rats were observed until spontaneous death and then necropsied. Histopathological examination of all organs and signs of toxicity or compound-related effects on survival were reported, and no compound-related histopathological lesions were observed. Thus, a NOAEL was found to be 41 mg/kg/day (10 g/0.35 kg/700 days assuming a body weight of 0.35 kg).

4. Reproductive Effects

- Pertinent data regarding the reproductive toxicity of naphthalene could not be located in the available literature.

5. Developmental Effects

- Matorova and Chetverikova (1981) treated pregnant outbred rats for 19 days with 0.075, 0.15 or 7 mg/kg/day naphthalene by gavage. Note that the highest dose administered is in question: tables in the paper list 70 mg/kg/day, instead of 7 mg/kg/day as cited in the text. The two higher doses resulted in observed developmental effects. However, due to lack of documentation, these results are viewed with caution.

- Plasterer et al. (1985) administered 300 mg/kg/day naphthalene by gavage to groups of 50 mated female CD-1 mice. Treatment began on gestation day 7 and continued for 8 consecutive days. Vehicle (corn oil) treated controls were maintained. Treatment was associated with maternal toxicity (increased mortality and reduced body weight gain) and fetotoxicity manifested as a reduced number of live young at birth, which the authors attributed to early fetal resorptions. Offspring were not examined for malformations.

- Hardin et al. (1981) reported the effects of fetal development after i.p. administration of 395 mg/kg naphthalene to adult female Sprague-Dawley rats (250 to 300 g) on days 1 through 15 of gestation. Control animals received 1 mg/kg corn oil, and control and treated litters were collected 1 day before parturition. There was no evidence of fetal or maternal toxicity.

- The administration to pregnant rabbits of 2-naphthol, a metabolite of naphthalene, has been associated with cataracts and evidence of retinal damage in the offspring (Van der Hoeve, 1913).

6. Mutagenicity

- Naphthalene was not active in reverse mutation assays in several strains of *Salmonella typhimurium* with or without metabolic activation (McCann et al., 1975; Seixas et al., 1982; NTP, 1982; Anderson and Styles, 1978; Hermann, 1981). Naphthalene was also not mutagenic in a *Salmonella* forward mutation assay with TM677 (Seixas et al., 1982; Kaden et al., 1979).

- Negative results were reported for the Rec assay in *Escherichia coli* in the presence or absence of an exogenous mammalian metabolism system (Mamber et al., 1983).

- Naphthalene has been tested for enhancement of cell transformation in rat and mouse embryo cells infected with leukemia virus (Freeman et al., 1973; Rhim et al., 1974) and in murine mammary gland organ cultures (Tonelli et al., 1979). No enhancement of transformation was observed.

- Naphthalene did not cause DNA damage in a rat hepatocyte alkaline elution assay (Sina et al., 1983).

7. Carcinogenicity

- No carcinogenic response was observed by Schmahl (1955) in rats given oral doses of 10 to 20 mg/day naphthalene, 6 days/week from day 100 to day 800 of age, or in rats given either subcutaneous (s.c.) or i.p.

injections of 20 mg naphthalene, 1 day/week for 40 weeks, and observed for the remainder of their lives.

- Knake (1956) reported increased incidence of lymphosarcomas in rats injected with 500 mg/kg coal tar naphthalene in sesame oil every 2 weeks for seven treatments; however, the injection site was painted with a carcinogen (carbolfuchsin) and the naphthalene was known to contain impurities. In another study, Knake (1956) painted mice with a 0.5% solution of coal tar naphthalene in benzene and reported an increased incidence of lymphatic leukemia. Again, the coadministration of carcinogenic impurities cannot be discounted.

- In a pulmonary adenoma induction test, inhalation of 0, 10 or 30 ppm (0, 52, 157 mg/m³) naphthalene by groups of 30 female strain A/J mice (6 to 8 weeks old) for 6 hours/day, 5 days/week for 6 months, did not produce a significant adenoma response (Adkins et al., 1986).

- Naphthalene is currently being evaluated by the National Toxicology Program (NTP) for carcinogenicity in mice by inhalation of doses of 10 or 30 ppm. Final results are not available (NTP, 1987).

IV. Quantification of Toxicological Effects

A. One-day Health Advisory

Sufficient data to derive a One-day HA for a 10-kg child are not available. Shopp et al. (1984) administered single gavage doses of 200, 400, 600, 800 or 1,000 mg/kg to CD-1 mice to determine the LD_{50} values. After a 14-day observation period, necropsy was performed on decedents and survivors. Ptosis and red ocular discharge were observed at ≥ 400 mg/kg doses. No adverse effects were reported at 200 mg/kg doses. Lezenius (1902) reported that naphthalene caused near blindness, cataract formation and optic atrophy in one human who ingested 5 g naphthalene in oil. Dreisbach and Robertson (1987) reported a fatal dose from oral exposure to be approximately 2 g. These doses are about 71 mg/kg and 28 mg/kg, respectively, for a 70-kg reference human. Because the 200 mg/kg NOAEL in mice reported by Shopp et al. (1984) is far greater than the 71 mg/kg and 28 mg/kg doses associated with adverse effects in humans, it cannot be used to derive a One-day HA. The Ten-day HA of 0.5 mg/L (described below) is recommended as a conservative estimate of the One-day HA for a 10-kg child.

B. Ten-day Health Advisory

Matorova and Chetverikova (1981) reported a NOAEL of 0.075 mg/kg/day for a 19-day exposure of pregnant outbred rats. Embryotoxicity and teratogenic effects were the end points examined; maternal toxicity was not evaluated. There was also a lack of documentation and some uncertainty regarding the reported results so that this study is not suitable for derivation of a Ten-day HA.

The 14-day gavage study by Shopp et al. (1984) in which CD-1 mice were exposed to naphthalene was considered as the basis for the 10-kg child Ten-day HA. In this study, increased mortality, decreased terminal body weights, altered organ weights and increased serum bilirubin were observed at the 267 mg/kg/day dose. No adverse effects were observed at the 27 or 53 mg/kg/day doses. Although the value of 53 mg/kg/day is the highest reported NOAEL, also note that Shopp et al. (1984) suggest that CD-1 mice are not an appropriate model for the pathological lesions most often found in humans (hemolytic anemia and cataract formation).

Lezenius (1902) reported that naphthalene caused near blindness, cataract formation and optic atrophy in one human who ingested 5 g naphthalene in oil; Dreisbach and Robertson (1987) reported a fatal dose from oral exposure to be approximately 2 g. These doses are about 71 mg/kg and 28 mg/kg, respectively, for a 70-kg reference human. Because the 53 mg/kg NOAEL in mice reported by Shopp et al. (1984) is close to the 71 mg/kg and 28 mg/kg doses associated with adverse effects in humans, an additional uncertainty factor is used to derive a Ten-day HA.

The Ten-day HA for the 10-kg child is calculated as follows:

$$\text{Ten-day HA} = \frac{(53 \text{ mg/kg/day}) (10 \text{ kg})}{(1,000) (1 \text{ L/day})} = 0.53 \text{ mg/L [rounded to } 0.5 \text{ mg/L (500 } \mu\text{g/L)]}$$

C. Longer-term Health Advisory

The 13 week gavage study by NTP (1980a), in which a NOAEL of 35.7 mg/kg/day (50 mg/kg for 5 days/week) was observed, has been selected to serve as the basis for the Longer-term HA. In this study F344 rats showed frank signs of toxicity, mortality and reduced blood hemoglobin and hematocrit at the 400 mg/kg dose. Mild kidney lesions were noted in male rats give the 200 mg/kg dose, and reduced rate of body weight gain was observed in female rats given the 100 mg/kg dose. No adverse effects were noted at 50 mg/kg (5 days/week). In an NTP (1980b) 13-week gavage study in which mice were treated 5 days/week with naphthalene, transient signs of toxicity that were observed at the 200 mg/kg dose were not attributed to treatment.

In a 90-day gavage study by Shopp et al. (1984), statistically significant reductions in brain, liver and spleen weights were observed in female CD-1 mice exposed to 133 mg/kg/day naphthalene. No organ weight changes were observed at the 5.3 or 53 mg/kg/day dose levels. Small changes in several serum chemistry parameters (i.e., blood protein, glucose, creatine, blood urea nitrogen, calcium ion), mostly at the middle and high doses, were considered to possibly be treatment related adverse effects. A statistically significant dose-related depression in hepatic aryl hydrocarbon hydroxylase was noted in all treated groups, but this may not represent an adverse effect. It should be noted that Shopp et al. (1984) suggest that CD-1 mice do not provide an appropriate model for the observed human effects of naphthalene, namely hemolytic anemia and cataract formation. A NOAEL could not be established from this study.

Lezenius (1902) reported that naphthalene caused near blindness, cataract formation and optic atrophy in one human who ingested 5 g naphthalene in oil; Dreisbach and Robertson (1987) reported a fatal dose from oral exposure to be approximately 2 g. These doses are about 71 mg/kg and 28 mg/kg, respectively, for a 70-kg reference human. Because the 35.7 mg/kg NOAEL in rats reported by NTP (1980a) is close to the 71 mg/kg and 28 mg/kg doses associated with adverse effects in humans, an additional uncertainty factor is used to derive a Longer-term HA.

The Longer-term HA for a 10-kg child is calculated as follows:

$$\text{Longer-term HA} = \frac{(35.7 \text{ mg/kg/day}) (10 \text{ kg})}{(1,000) (1 \text{ L/day})} = 0.357 \text{ mg/L [rounded to } 0.4 \text{ mg/L } (400 \text{ } \mu g/L)]$$

The Longer-term HA for a 70-kg adult is calculated as follows:

$$\text{Longer-term HA} = \frac{(35.7 \text{ mg/kg/day}) (70 \text{ kg})}{(1,000) (2 \text{ L/day})} = 1.249 \text{ mg/L [rounded to } 1 \text{ mg/L } (1,000 \text{ } \mu g/L)]$$

D. Lifetime Health Advisory

The 13 week NTP (1980a) gavage study in which a NOAEL of 35.7 mg/kg/day (50 mg/kg/day for 5 days/week) was used to derive the Longer-term HA has also been selected to serve as the basis for the Lifetime HA. The chronic feeding study by Schmahl (1955) in which 100-day-old rats were fed a total of 10 g naphthalene over a 700-day period and observed for their natural lifespan is a supporting study for the Lifetime HA. In the Schmahl (1955) study, histopathological examination of all organs and signs of toxicity or compound-related effects on survival were reported, no compound-related histopathological lesions were observed, and a NOAEL was found to be 41 mg/kg/day (10 g/0.35 kg/700 days assuming a body weight of 0.35 kg).

Lezenius (1902) reported that naphthalene caused near blindness, cataract formation and optic atrophy in one human who ingested 5 g naphthalene in oil; Dreisbach and Robertson (1987) reported a fatal dose from oral exposure to be approximately 2 g. These doses are about 71 mg/kg and 28 mg/kg, respectively, for a 70-kg reference human. Because the 35.7 mg/kg NOAEL in rats reported by NTP (1980a) is close to the 71 mg/kg and 28 mg/kg doses associated with adverse effects in humans, an additional uncertainty factor is used to derive a Lifetime HA. Using the NTP (1980a) study, the DWEL is derived as follows:

Step 1: Determination of the Reference Dose (RfD)

$$RfD = \frac{(35.7 \text{ mg/kg/day})}{(10,000)} = 0.00357 \text{ mg/kg/day (rounded to } 0.004 \text{ mg/kg/day)}$$

Step 2: Determination of the Drinking Water Equivalent Level (DWEL)

$$DWEL = \frac{(0.00357 \text{ mg/kg/day}) (70 \text{ kg})}{(2 \text{ L/day})} = 0.1249 \text{ mg/L [rounded to } 0.1 \text{ mg/L (100 } \mu g/L)]$$

Step 3: Determination of the Lifetime Health Advisory

$$\text{Lifetime HA} = (0.1249 \text{ mg/L}) (20 \text{ \%}) = 0.02498 \text{ mg/L [rounded to } 0.02 \text{ mg/L (20 } \mu g/L)]$$

D. Evaluation of Carcinogenic Potential

- Naphthalene was not carcinogenic to BDI or BDIII rats treated orally with 10 to 20 mg/day, 6 days/week, from 100 to 800 days of age and observed for their natural lifespan (Schmahl, 1955). Total dose was 10 g or 41 mg/kg/day. Data from the 13-week study by NTP (1980a) indicate that this dose is not the maximum tolerated dose. An inhalation study in mice is currently being conducted by NTP (1987).

- IARC has not evaluated the carcinogenic potential of naphthalene, but the U.S. EPA (1986a) classified naphthalene in IARC Group 3.

- Applying the criteria described in the U.S. EPA's Guidelines for Carcinogen Risk Assessment (U.S. EPA, 1986b), naphthalene may be classified in Group D: not classifiable. This category is for agents for which there is inadequate human or animal evidence of carcinogenicity or for which no data are available.

V. Other Criteria, Guidance and Standards

- ACGIH (1980) recommended a TWA-TLV of 10 ppm (50 mg/m³) and a Short-term Exposure Level (STEL) of 15 ppm (75 mg/m³) to prevent ocular effects but not necessarily blood changes in hypersensitive individuals. The OSHA PEL for naphthalene is also 10 ppm (50 mg/m³) (CFR, 1985).

VI. Analytical Methods

- Analysis of naphthalene is by a solvent extraction gas chromatographic procedure used for the determination of polynuclear aromatic hydrocarbons in water samples (U.S. EPA, 1984a). This method requires the extraction of a 1 liter sample with methylene chloride using a separatory funnel. The methylene chloride extract is dried and concentrated to a volume of ≤ 10 mL. The extract is then separated by high pressure liquid chromatography (HPLC) or gas chromatography (GC). The reported method detection limit for naphthalene using HPLC with an ultraviolet detector has been determined to be 1.8 μg/L. Confirmatory analysis for this contaminant is by gas chromatography/mass spectrometry (U.S. EPA, 1984b). The detection limit for confirmation by mass spectrometry is 1.6 μg/L.

VII. Treatment Technologies

- Treatment technologies which will remove naphthalene from drinking water include adsorption, aeration, reverse osmosis, steam stripping and potassium ferrate oxidation-coagulation. No data were found for the removal of naphthalene from drinking water by conventional treatment or reverse osmosis.

- Walters and Luthy (1984) developed adsorption isotherm data for naphthalene from water onto Filtrasorb 400 granular activated carbon (GAC). The following Freundlich parameters were reported:

$$K = 132$$
$$1/n = 0.42$$

For initial concentrations of 10, 100 and 3,000 μg/L, equilibrium adsorption capacities of 40, 100 and 800 mg/g, respectively, were reported. Dobbs and Cohen (1980) performed isotherm tests on the removal of naphthalene from drinking water by GAC. The following Freundlich parameters were reported:

$$K = 132$$
$$1/n = 0.42$$

For initial naphthalene concentrations of 10, 100 and 1,000 μg/L, equilibrium adsorption capacities of 1, 50 and 132 mg/g, respectively, were obtained.

- Singley et al. (1981) reported on the removal of naphthalene with powdered activated carbon (PAC), applied at a dosage of 7.1 mg/L, at the Sunny Isles WTP in North Miami Beach, FL. For a naphthalene influent concentration of 0.006 μg/L, 75% removal was obtained.

- Mumford and Schnoor (1982) performed packed column aeration pilot studies with air-to-water ratios between 25:1 and 200:1 with 1/4-inch ceramic berl saddles as the packing material. For a packing height of 0.457 m, 36 to 91% of the naphthalene was removed. Higher percent removals reflected higher air-to-water ratios.

- Reinhard et al. (1986) reported on the removal of naphthalene from water by reverse osmosis at Water Factory 21 in Orange County, CA. A full-scale unit with a cellulose acetate membrane operating at 30% recovery achieved a 54% naphthalene reduction. Two pilot-scale units with polyamide membranes operating at 37 and 72% recovery achieved 80 and 30% removal of naphthalene, respectively.

- Hwang and Fahrenthold (1980) reported that steam stripping of water saturated with 30 mg/L of naphthalene achieved 97% removal in a four-stage column which required 0.01 kg of steam per kg of feed.

- DeLuca et al. (1983) studied the removal of organic priority pollutants by potassium ferrate oxidation-coagulation. Potassium ferrate oxidation-coagulation with paddle flocculation and with gas (N_2) flocculation both achieved 100% removal of naphthalene.

VIII. References

ACGIH. 1980. American Conference of Governmental Hygienists. Documentation of threshold limit values, 4th ed. (Includes supplemental documentation through 1984.) Cincinnati, OH: ACGIH. p. 293.

Adkins, B., Jr., E.W. Van Stee, J.E. Simmons and S.L. Eustis. 1986. Oncogenic response of strain A/J mice to inhaled chemicals. J. Toxicol. Environ. Health. 17:311–322.

Amoore, J.E. and E. Hautala. 1983. Odor as an aid to chemical safety: odor thresholds compared with threshold limit values and volatilities for 214 industrial chemicals in air and water dilution. J. Appl. Toxicol. 3(6):272–290.

Anderson, D. and J.A. Styles. 1978. An evaluation of 6 short-term tests for detecting organic chemical carcinogens. Appendix 2. The bacterial mutation test. Br. J. Cancer. 37:924–930.

Anziulewicz, J.A., H.J. Dick and E.E. Chiaralli. 1959. Transplacental naphthalene poisoning. Am. J. Obstet. Gynecol. 78:519–521. (Cited in U.S. EPA, 1980.)

Benoit, F.M., G.L. Lebel and D.T. Williams. 1979. The determination of polycyclic aromatic hydrocarbons at the ng/L level in Ottawa tap water. Int. J. Environ. Anal. Chem. 6:227–287.

Boyd, D.R., J.W. Daly and D.M. Jerina. 1972. Rearrangement of $(1\text{-}H^2)$ and $(2\text{-}H^2)$ naphthalene-1,2-oxides to 1-naphthol. Mechanism of the NIH shift. Biochem. J. 11:1961–1966. (Cited in U.S. EPA, 1980.)

Callahan, M.A., D.J. Ehreth and P.L. Levins. 1979. Sources of toxic pollutants found in influent to sewage treatment plants. Proc. Natl. Conf. Munic. Sludge Manage. 8:55–61.

Chemline. 1988. Online. Bethesda, MD: National Library of Medicine.

Chen, K. and H.W. Dorough. 1979. Glutathione and mercapturic acid conjugations in the metabolism of naphthalene and 1-naphthyl-N-methylcarbamate (carbaryl). Drug Chem. Toxicol. 2(4):331–354.

CFR. 1985. Code of Federal Regulations. OSHA occupational standards. 29 CFR:1910.1000.

Cole, R.H., R.E. Frederick, R.P. Healy and R.G. Rolan. 1984. Preliminary findings of the priority pollutant monitoring project of the Nationwide Urban Runoff Program. J. Water Pollut. Control Fed. 56(7):898–908.

DeLuca, S.J., A.C. Chao and C. Smallwood. 1983. Removal of organic priority pollutants by oxidation-coagulation. J. Environ. Engineer. 109(1):36–46.

Dobbs, R.A. and J.M. Cohen. 1980. Carbon absorption of isotherms for toxic organics. Cincinnati, OH: Office of Research and Development, MERL, Wastewater Research Division. EPA 600/8-80-023.

Dreisbach, R.H. and W.O. Robertson. 1987. Handbook of poisoning: prevention, diagnosis and treatment, 12th ed. Norwalk, CT: Appleton and Lange. p. 194.

Eisele, G.R. 1985. Naphthalene distribution in tissues of laying pullets, swine, and dairy cattle. Bull. Environ. Contam. Toxicol. 34(4):549–556.

Freeman, A.E., E.K. Weisburger, J.H. Weisburger, R.G. Wolford, J.M. Maryak and R.J. Huebner. 1973. Transformation of cell cultures as an indication of the carcinogenic potential of chemicals. J. Natl. Cancer Inst. 51:799–807. (Cited in U.S. EPA, 1980.)

Gaines, T.B. 1969. Acute toxicity of pesticides. Toxicol. Appl. Pharmacol. 14:515.

Galloway, W.B., J.L. Lake, D.K. Phelps et al. 1983. The mussel watch: intercomparison of trace level constituent determinations. Environ. Toxicol. Chem. 2:395–410.

Garten, C.T., Jr. and J.R. Trabalka. 1983. Evaluation of models for predicting terrestrial food chain behavior of xenobiotics. Environ. Sci. Technol. 17:590–595. (Cited in U.S. EPA, 1986a.)

Ghetti, G. and L. Mariani. 1956. Eye changes due to naphthalene. Med. Lav. 47:533–538. (Cited in U.S. EPA, 1980.)

Hansch, C. and A.J. Leo. 1985. Medchem Project. Issue No. 26. Claremont, CA: Pomona College. (Cited in U.S. EPA, 1986a.)

Hardin, B.D., G.P. Bond, M.R. Sikov, F.D. Andrew, R.P. Beliles and R.W. Niemeier. 1981. Testing of selected workplace chemicals for teratogenic potential. Scan. J. Work Environ. Health. 7(Suppl. 4): 66–75.

Hermann, M. 1981. Synergistic effects of individual polycyclic aromatic hydrocarbons on the mutagenicity of their mixtures. Mutat. Res. 90(4):399–409.

Hine, J. and P.K. Mookerjee. 1975. The intrinsic hydrophilic character of organic compounds. Correlations in terms of structural contributions. J. Org. Chem. 40:292–298.

Hwang, S.T. and P. Fahrenthold. 1980. Treatability of the organic priority pollutants by steam stripping. AICHE Symposium Series. 76(197):37–60.

Jarke, F.H., A. Dravnieks and S.M. Gordon. 1981. Organic contaminants in indoor air and their relation to outdoor contaminants. ASHRAE Trans. 87:153–166.

Jerina, D.M., J.W. Daly, B. Witkop, P. Zaltzman-Nirenberg and S. Udenfriend. 1970. 1,2-Naphthalene oxide as an intermediate in the microsomal hydroxylation of naphthalene. Biochemistry. 9(1):147–155.

Jungclaus, G.A., V. Lopez-Avula and R.A. Hites. 1978. Organic compounds in an industrial wastewater: a case study of their environmental impact. Environ. Sci. Technol. 12(1):88–96.

Jurg, W.A., W.F. Spencer and W.J. Farmer. 1984. Behavior assessment model for trace organics in soil. III. Application of screening model. J. Environ. Qual. 13:573–579. (Cited in U.S. EPA, 1986a.)

Kaden, D.A., R.A. Hites and W.G. Thilly. 1979. Mutagenicity of soot and associated polycyclic aromatic hydrocarbons to *Salmonella typhimurium*. Cancer Res. 39(10):4152–4159.

Kenaga, E.E. 1980. Predicted bioconcentration factors and soil sorption coefficients of pesticides and other chemicals. Ecotoxicol. Environ. Saf. 4:26–38.

Knake, E. 1956. *Uber schwache geschwulsterzengende Wirkung von Naphtha-*

lin und Benzol. Virchows Archiv. Pathol. Anat. Physiol. 329:141–146. (Ger.) (Cited in U.S. EPA, 1980.)

Kveseth, K., B. Sortland and T. Bokn. 1982. Polycyclic aromatic hydrocarbons in sewage, mussels and tap water. Chemosphere. 11(7):623–639.

Lee, R.F., W.S. Gardner, J.W. Anderson, J.W. Blaylock and J. Barwell-Clarke. 1978. Fate of polycyclic aromatic hydrocarbons in controlled ecosystem enclosures. Environ. Sci. Technol. 12(7):832–838.

Lezenius, A. 1902. *Ein fall von naphthalinkatarakt am mansonen*. Klin. Mbl. Augenheilk. 40:129. (Ger.) (Cited in U.S. EPA, 1980.)

Mackay, D. and P.J. Leinonen. 1975. Rate of evaporation of low-solubility contaminants from water bodies to atmosphere. Environ. Sci. Technol. 9(13):1178–1180.

Mamber, S.W., V. Bryson and S.E. Katz. 1983. The *Escherichia coli* WP2/WP100 rec assay for detection of potential chemical carcinogens. Mutat. Res. 119:135–144.

Matorova, N.N. and O.N. Chetverikova. 1981. Embryotoxic and teratogenic effect of naphthalene and chloronaphthalene. Sb. Nauch. Tr. VNII Gigieny. 12:62–65. (English translation)

McCann, J., E. Choi, E. Yamasaki and B.N. Ames. 1975. Detection of carcinogens as mutagens in the *Salmonella*/microsome test: assay of 300 chemicals. Proc. Natl. Acad. Sci. USA. 72:5135–5139.

Mumford, R.L. and J.L. Schnoor. 1982. Air stripping of volatile organics in water. Proc. 1982 AWWA Annual Conference. pp. 601–617.

NIOSH. 1977. National Institute for Occupational Safety and Health. 1977. Registry of toxic effects of chemical substances, Vol. II. NIOSH Publ. No. 78-104-B. U.S. DHEW. (Cited in U.S. EPA, 1980.)

NTP. 1980a. National Toxicology Program. Unpublished subchronic toxicity study: naphthalene (C52904), Fischer 344 rats. Prepared by Battelle's Columbus Laboratories. Subcontract No. 76-34-106002. March 4.

NTP. 1980b. National Toxicology Program. Unpublished subchronic toxicity study: naphthalene (C52904), B6C3F$_1$ mice. Prepared by Battelle's Columbus Laboratories. Subcontract No. 76-34-106002. March 4.

NTP. 1982. National Toxicology Program. Technical Bulletin. Issue 6, January. DHHS.

NTP. 1987. National Toxicology Program. Toxicology Research and Testing Program. Management Status Report. Research Triangle Park, NC: NTP. July 11.

Pellizzari, E.D., T.D. Hartwell, B.S.H. Harris, III, R.D. Waddell, D.A. Whitaker and M.D. Erickson. 1982. Purgeable organic compounds in mother's milk. Bull. Environ. Contam. Toxicol. 28:322–328.

Plasterer, M.R., W.S. Bradshaw, G.M. Booth, M.W. Carger, R.L. Schuler and B.D. Hardin. 1985. Developmental toxicity of nine selected compounds following prenatal exposure in the mouse: naphthalene, p-nitrophenol, sodium selenite, dimethyl phthalate, ethylenethiourea, and four glycol ether derivatives. J. Toxicol. Environ. Health. 15(1):25–38.

Reinhard, M., N.L. Goodman, P.L. McCarty and D.G. Argo. 1986. Removing trace organics by reverse osmosis using cellulose acetate and polyamide membranes. JAWWA. 78(4):163–174.

Rhim, J.S., D.K. Park, E.K. Weisburger and J.H. Weisburger. 1974. Evaluation of an *in vitro* assay system for carcinogens based on prior infection of rodent cells with nontransforming RNA tumor virus. J. Natl. Cancer Inst. 52:1167–1173. (Cited in U.S. EPA, 1980.)

Rodgers, J.H., Jr., K.L. Dickson, F.Y. Saleh and C.A. Staples. 1983. Use of microcosms to study transport, transformation and fate of organics in aquatic systems. Environ. Toxicol. Chem. 2:155–167.

Sandmeyer, E.E. 1981. Aromatic hydrocarbons. In: G.D. Clayton and F.E. Clayton, eds. Patty's Industrial Hygiene and Toxicology, Vol. 2B, 34th ed. New York, NY: John Wiley and Sons, Inc. pp. 3253–3431.

Scheiman, M.A., R.A. Saunders and F.E. Saalfeld. 1974. Organic contaminants in the District of Columbia water supply. Biomed. Mass Spectrom. 4:209–211.

Schmahl, D. 1955. Testing of naphthalene and anthracene as carcinogenic agents in the rat. Z. Krebsforsch. 60:697–710. (Ger.) (Partial translation)

Seixas, G.M., B.M. Andon, P.G. Hollingshead and W.B. Thilly. 1982. The aza-arenes as mutagens for *Salmonella typhimurium*. Mutat. Res. 102:201–212.

Sheldon, L.S. and R.A. Hites. 1978. Organic compounds in the Delaware River. Environ. Sci. Technol. 12(10):1188–1194. (Cited in U.S. EPA, 1986a.)

Sheldon, L.S. and R.A. Hites. 1979. Sources and movement of organic chemicals in the Delaware River. Environ. Sci. Technol. 13(5):574–579. (Cited in U.S. EPA, 1986a.)

Shinohara, R., A. Kido, S. Eto, T. Hori, M. Koga and T. Akiyama. 1981. Identification and determination of trace organic substances in tap water by computerized gas chromatography-mass spectrometry and mass fragmentography. Water Res. 15:535–542.

Shopp, G.M., K.L. White, Jr., M.P. Holsapple et al. 1984. Naphthalene toxicity in CD-1 mice: general toxicology and immunotoxicology. Fund. Appl. Toxicol. 4(3, Pt. 1):406–419.

Sina, J.F., C.L. Bean, G.R. Dysart, V.I. Taylor and M.O. Bradley. 1983. Evaluation of the alkaline elution/rat hepatocyte assay as a predictor of carcinogenic/mutagenic potential. Mut. Res. 113:357–391.

Singley, J.E., B.A. Beaudet and A.L. Ervin. 1981. Use of powdered activated carbon for removal of specific organic compounds. AWWA Seminar Proceedings–Organic Chemical Contamination in Ground Water.

Southworth, G.R. 1979. The role of volatilization in removing polycyclic aromatic hydrocarbons from aquatic environments. Bull. Environ. Contam. Toxicol. 21:507–514.

Staples, C.A., A.F. Werner and T.J. Hoogheem. 1985. Assessment of priority pollutant concentrations in the United States using STORET data base. Environ. Toxicol. Chem. 4:131–142.

Summer, K.H., K. Rozman, F. Coulston and H. Greim. 1979. Urinary excretion of mercapturic acids in chimpanzees and rats. Toxicol. Appl. Pharmacol. 50(2):207–212.

Tonelli, Q., R.P. Custer and S. Sorof. 1979. Transformation of cultured mouse mammary glands by aromatic amines and amides and their derivatives. Cancer Res. 39:1784–1792. (Cited in U.S. EPA, 1980.)

Union Carbide Corp. 1968. Naphthalene safety data sheet. New York. (Cited in U.S. EPA, 1980.)

U.S. EPA. 1980. U.S. Environmental Protection Agency. Ambient water quality criteria document for naphthalene. Prepared by the Office of Health and Environmental Assessment, Environmental Criteria and Assessment Office, Cincinnati, OH for the Office of Water Regulations and Standards, Criteria and Standards Division, Washington, DC. EPA 440/5-80-059. NTIS PBB1-117707.

U.S. EPA. 1982. U.S. Environmental Protection Agency. Volatile organic chemicals in the atmosphere: an assessment of available data. Final report. Contract No. 68-02-3452. Research Triangle Park, NC: Environmental Science Research Laboratory, ORD. (Cited in U.S. EPA, 1986a.)

U.S. EPA. 1984a. U.S. Environmental Protection Agency. U.S. EPA Method 610–Polynuclear aromatic hydrocarbons. *Fed. Reg.* 45(209):112–120. October 26.

U.S. EPA. 1984b. U.S. Environmental Protection Agency. U.S. EPA Method 625–Base/neutral and acids. *Fed. Reg.* 49(209):153–174. October 26.

U.S. EPA. 1986a. U.S. Environmental Protection Agency. Health and environmental effects profile for naphthalene. Prepared by the Office of Health and Environmental Assessment, Environmental Criteria and Assessment Office, Cincinnati, OH for the Office of Solid Waste and Emergency Response, Washington, DC.

U.S. EPA. 1986b. U.S. Environmental Protection Agency. Guidelines for carcinogen risk assessment. *Fed. Reg.* 51(185):33992–34003. September 24.

Valaes, T., S.A. Doxiadis and P. Fessas. 1963. Acute hemolysis due to naphthalene inhalation. J. Pediat. 63:904–915. (Cited in U.S. EPA, 1980.)

Van der Hoeve, J. 1906. *Chorioretinitis beim menschen durch die einwirking von naphthalin*. Arch. Augenheiklk. 56:259. (Ger.) (Cited in U.S. EPA, 1980.)

Van der Hoeve, J. 1913. *Wirkung von naphthol auf die augen van menschen, tieren, und auf fatale augen*. Graele Arch. Ophthal. 85:305. (Ger.) (Cited in U.S. EPA, 1980.)

Van Heyningen, R. and A. Pirie. 1976. Naphthalene cataract in pigmented and albino rabbits. Exp. Eye Res. 22(4):393–394.

Verschueren, K. 1983. Handbook of Environmental Data on Organic Chemicals, 2nd ed. New York, NY: Van Nostrand Reinhold Co. pp. 779–898.

Walters, R. W. and R.G. Luthy. 1984. Equilibrium adsorption of polycyclic aromatic hydrocarbons from water onto activated carbon. Environ. Sci. Tech. 18(6):395–403.

Windholz, M., ed. 1983. The Merck Index — An Encyclopedia of Chemicals, Drugs and Biologicals, 10th ed. Rahway, NJ: Merck and Co., Inc. p. 914.

Zinkham, W.J. and B. Childs. 1958. A defect of glutathione metabolism in erythrocytes from patients with a naphthalene-induced hemolytic anemia. Pediatrics. 22:461–470. (Cited in U.S. EPA, 1980.)

Zoeteman, B.C.J., K. Harsen, J.B.H.J. Linders, C.F.H. Morra and W. Slooff. 1980. Persistent organic pollutants in river water and ground water of the Netherlands. Chemosphere. 9:231–249.

Zoeteman, B.C.J., E. Degreef and F.J.H. Brinkman. 1981. Persistence of organic contaminants in ground water, lessons from soil pollution incidents in the Netherlands. Sci. Total Environ. 21:187–202. (Cited in U.S. EPA, 1986a.)

Zuelzer, W.W. and L. Apt. 1949. Acute hemolytic anemia due to naphthalene poisoning. J. Am. Med. Assoc. 141:185–190. (Cited in U.S. EPA, 1980.)